T0181377

Intelligent Systems Reference Library

Volume 117

Series editors

Janusz Kacprzyk, Polish Academy of Sciences, Warsaw, Poland
e-mail: kacprzyk@ibspan.waw.pl

Lakhmi C. Jain, University of Canberra, Canberra, Australia;
Bournemouth University, UK;
KES International, UK
e-mails: jainlc2002@yahoo.co.uk; Lakhmi.Jain@canberra.edu.au
URL: http://www.kesinternational.org/organisation.php

About this Series

The aim of this series is to publish a Reference Library, including novel advances and developments in all aspects of Intelligent Systems in an easily accessible and well structured form. The series includes reference works, handbooks, compendia, textbooks, well-structured monographs, dictionaries, and encyclopedias. It contains well integrated knowledge and current information in the field of Intelligent Systems. The series covers the theory, applications, and design methods of Intelligent Systems. Virtually all disciplines such as engineering, computer science, avionics, business, e-commerce, environment, healthcare, physics and life science are included.

More information about this series at http://www.springer.com/series/8578

Diego Oliva · Erik Cuevas

Advances and Applications of Optimised Algorithms in Image Processing

Springer

Diego Oliva
Departamento de Electrónica, CUCEI
Universidad de Guadalajara
Guadalajara, Jalisco
Mexico

and

Tecnológico de Monterrey, Campus
  Guadalajara
Zapopan, Jalisco
Mexico

Erik Cuevas
Departamento de Electrónica, CUCEI
Universidad de Guadalajara
Guadalajara, Jalisco
Mexico

ISSN 1868-4394   ISSN 1868-4408   (electronic)
Intelligent Systems Reference Library
ISBN 978-3-319-83969-1   ISBN 978-3-319-48550-8   (eBook)
DOI 10.1007/978-3-319-48550-8

© Springer International Publishing AG 2017
Softcover reprint of the hardcover 1st edition 2016
This work is subject to copyright. All rights are reserved by the Publisher, whether the whole or part
of the material is concerned, specifically the rights of translation, reprinting, reuse of illustrations,
recitation, broadcasting, reproduction on microfilms or in any other physical way, and transmission
or information storage and retrieval, electronic adaptation, computer software, or by similar or dissimilar
methodology now known or hereafter developed.
The use of general descriptive names, registered names, trademarks, service marks, etc. in this
publication does not imply, even in the absence of a specific statement, that such names are exempt from
the relevant protective laws and regulations and therefore free for general use.
The publisher, the authors and the editors are safe to assume that the advice and information in this
book are believed to be true and accurate at the date of publication. Neither the publisher nor the
authors or the editors give a warranty, express or implied, with respect to the material contained herein or
for any errors or omissions that may have been made.

Printed on acid-free paper

This Springer imprint is published by Springer Nature
The registered company is Springer International Publishing AG
The registered company address is: Gewerbestrasse 11, 6330 Cham, Switzerland

To my family and Gosia Kijak, you are always my support

Foreword

This book brings together and explores possibilities for combining image processing and artificial intelligence, both focused on machine learning and optimization, two relevant areas and fields in computer science. Most books have been proposed about the major topics separately, but not in conjunction, giving it a special interest. The problems addressed and described in the different chapters were selected in order to demonstrate the capabilities of optimization and machine learning to solve different issues in image processing. These problems were selected considering the degree of relevance in the field providing important cues on particular applications domains. The topics include the study of different methods for image segmentation, and more specifically detection of geometrical shapes and object recognition, where their applications in medical image processing, based on the modification of optimization algorithms with machine learning techniques, provide a new point of view. In short, the book was intended with the purpose and motivation to show that optimization and machine learning main topics are attractive alternatives for image processing technique taking advantage over other existing strategies. Complex tasks can be addressed under these approaches providing new solutions or improving the existing ones thanks to the required foundation for solving problems in specific areas and applications.

Unlike other existing books in similar areas, the book proposed here introduces to the reader the new trends using optimization approaches about the use of optimization and machine learning techniques applied to image processing. Moreover, each chapter includes comparisons and updated references that support the results obtained by the proposed approaches, at the same time that provides the reader a practical guide to go to the reference sources.

The book was designed for graduate and postgraduate education, where students can find support for reinforcing or as the basis for their consolidation or deepening of knowledge, and for researchers. Also teachers can find support for the teaching process in the areas involving machine vision or as examples related to main techniques addressed. Additionally, professionals who want to learn and explore the advances on concepts and implementation of optimization and learning-based algorithms applied image processing find in this book an excellent guide for such purpose.

The content of this book has been organized considering an introduction to machine learning an optimization. After each chapter addresses and solves selected problems in image processing. In this regard, Chaps. 1 and 2 provides respectively introductions to machine learning and optimization, where the basic and main concepts related to image processing are addressed. Chapter 3, describes the electromagnetism-like optimization (EMO) algorithm, where the appropriate modifications are addressed to work properly in image processing. Moreover, its advantages and shortcomings are also explored. Chapter 4 addresses the digital image segmentation as an optimization problem. It explains how the image segmentation is treated as an optimization problem using different objective functions. Template matching using a physical inspired algorithm is addressed in Chap. 5, where indeed, template matching is considered as an optimization problem, based on a modification of EMO and considering the use of a memory to reduce the number of call functions. Chapter 6 addresses the detection of circular shapes problem in digital images, and again focused as an optimization problem. A practical medical application is proposed in Chap. 7, where blood cell segmentation by circle detection is the problem to be solved. This chapter introduces a new objective function to measure the match between the proposed solutions and the blood cells contained in the images. Finally, Chap. 8 proposes an improvement EMO applying the concept of opposition-based electromagnetism-like optimization. This chapter analyzes a modification of EMO used as a machine learning technique to improve its performance. An important advantage of this structure is that each chapter could be read separately. Although all chapters are interconnected, Chap. 3 serves as the basis for some of them.

The concise comprehensive book on the topics addressed makes this work an important reference in image processing, which is an important area where a significant number of technologies are continuously emerging and sometimes untenable and scattered along the literature. Therefore, congratulations to authors for their diligence, oversight and dedication for assembling the topics addressed in the book. The computer vision community will be very grateful for this well-done work.

July 2016   Gonzalo Pajares
                                                    Universidad Complutense de Madrid

Preface

The use of cameras to obtain images or videos from the environment has been extended in the last years. Now these sensors are present in our lives, from cell phones to industrial, surveillance and medical applications. The tendency is to have automatic applications that can analyze the images obtained with the cameras. Such applications involve the use of image processing algorithms.

Image processing is a field in which the environment is analyzed using samples taken with a camera. The idea is to extract features that permit the identification of the objects contained in the image. To achieve this goal is necessary applying different operators that allow a correct analysis of a scene. Most of these operations are computationally expensive. On the other hand, optimization approaches are extensively used in different areas of engineering. They are used to explore complex search spaces and obtain the most appropriate solutions using an objective function. This book presents a study the uses of optimization algorithms in complex problems of image processing. The selected problems explore areas from the theory of image segmentation to the detection of complex objects in medical images. The concepts of machine learning and optimization are analyzed to provide an overview of the application of these tools in image processing.

The aim of this book is to present a study of the use of new tendencies to solve image processing problems. When we start working on those topics almost ten years ago, the related information was sparse. Now we realize that the researchers were divided and closed in their fields. On the other hand, the use of cameras was not popular then. This book presents in a practical way the task to adapt the traditional methods of a specific field to be solved using modern optimization algorithms. Moreover, in our study we notice that optimization algorithm could also be modified and hybridized with machine learning techniques. Such modifications are also included in some chapters. The reader could see that our goal is to show that exist a natural link between the image processing and optimization. To achieve this objective, the first three chapters introduce the concepts of machine learning, optimization and the optimization technique used to solve the problems. The structure of the rest of the sections is to first present an introduction to the problem to be solved and explain the basic ideas and concepts about the implementations.

The book was planned considering that, the readers could be students, researchers expert in the field and practitioners that are not completely involved with the topics.

This book has been structured so that each chapter can be read independently from the others. Chapter 1 describes the machine learning (ML). This chapter concentrates on elementary concepts of machine learning. Chapter 2 explains the theory related with global optimization (GO). Readers that are familiar with those topics may wish to skip these chapters.

In Chap. 3 the electromagnetism-like optimization (EMO) algorithm is introduced as a tool to solve complex optimization problems. The theory of physics behind the EMO operators is explained. Moreover, their pros and cons are widely analyzed, including some of the most significant modifications.

Chapter 4 presents three alternative methodologies for image segmentation considering different objective functions. The EMO algorithm is used to find the best thresholds that can segment the histogram of a digital image.

In Chap. 5 the problem template matching is introduced that consists in the detection of objects in an image using a template. Here the EMO algorithm optimizes an objective function. Moreover, improvements to reduce the number of evaluations and the convergence velocity are also explained.

Continuing with the object detection, Chap. 6 shows how EMO algorithm can be applied to detect circular shapes embedded in digital images. Meanwhile, in Chap. 7 a modified objective function is used to identify white blood cells in medical images using EMO.

Chapter 8 shows how a machine learning technique could improve the performance of an optimization algorithm without affecting its main features such as accuracy or convergence.

Writing this book was a very rewarding experience where many people were involved. We acknowledge Dr. Gonzalo Pajares for always being available to help us. We express our gratitude to Prof. Lakhmi Jain, who so warmly sustained this project. Acknowledgements also go to Dr. Thomas Ditzinger, who so kindly agreed to its appearance.

Finally, it is necessary to mention that this book is a small piece in the puzzles of image processing and optimization. We would like to encourage the reader to explore and expand the knowledge in order create their own implementations according their own necessities.

Zapopan, Mexico   Diego Oliva
Guadalajara, Mexico   Erik Cuevas
July 2016

Contents

Chapter 1
An Introduction to Machine Learning

1.1 Introduction

We already are in the era of big data. The overall amount of data is steadily growing. There are about one trillion of web pages; one hour of video is uploaded to YouTube every second, amounting to 10 years of content every day. Banks handle more than 1 M transactions per hour and has databases containing more than 2.5 petabytes (2.5×10^{15}) of information; and so on [1].

In general, we define machine learning as a set of methods that can automatically detect patterns in data, and then use the uncovered patterns to predict future data, or to perform other kinds of decision making under uncertainty. Learning means that novel knowledge is generated from observations and that this knowledge is used to achieve defined objectives. Data itself is already knowledge. But for certain applications and for human understanding, large data sets cannot directly be applied in their raw form. Learning from data means that new condensed knowledge is extracted from the large amount of information [2].

Some typical machine learning problems include, for example in bioinformatics, the analysis of large genome data sets to detect illnesses and for the development of drugs. In economics, the study of large data sets of market data can improve the behavior of decision makers. Prediction and inference can help to improve planning strategies for efficient market behavior. The analysis of share markets and stock time series can be used to learn models that allow the prediction of future developments. There are thousands of further examples that require the development of efficient data mining and machine learning techniques. Machine learning tasks vary in various kinds of ways, e.g., the type of learning task, the number of patterns, and their size [2].

© Springer International Publishing AG 2017
D. Oliva and E. Cuevas, *Advances and Applications of Optimised Algorithms in Image Processing*, Intelligent Systems Reference Library 117,
DOI 10.1007/978-3-319-48550-8_1

1.2   Typed of Machine Learning Strategies

The Machine learning methods are usually divided into three main types: supervised, unsupervised and reinforcement learning [3]. In the predictive or supervised learning approach, the goal is to learn a mapping from inputs $\mathbf{x}$ to outputs y, given a labeled set of input-output pairs $\mathbf{D} = \{(\mathbf{x}_i, y_i)\}_{i=1}^{N}, \mathbf{x}_i = \left(x_i^1, \ldots, x_i^d\right)$. Here $\mathbf{D}$ is called the training data set, and N represents the number of training examples.

In the simplest formulation, each training vector $\mathbf{x}$ is a d-dimensional vector, where each dimension represents a feature or attribute of $\mathbf{x}$. Similarly, y_i symbolizes the category assigned to $\mathbf{x}_i$. Such categories integrate a set defined as $y_i \in \{1, \ldots, C\}$. When y_i is categorical, the problem is known as classification and when y_i is real-valued, the problem is known as regression. Figure 1.1 shows a schematic representation of the supervised learning.

The second main method of machine learning is the unsupervised learning. In unsupervised learning, it is only necessary to provide the data $\mathbf{D} = \{\mathbf{x}_i\}_{i=1}^{N}$. Therefore, the objective of an unsupervised algorithm is to automatically find patterns from the data, which are not initially apparent. This process is sometimes called knowledge discovery. Under such conditions, this process is a much less well-defined problem, since we are not told what kinds of patterns to look for, and there is no obvious error metric to use (unlike supervised learning, where we can compare our prediction of y_i for a given $\mathbf{x}_i$ to the observed value). Figure 1.2 illustrate the process of unsupervised learning. In the figure, data are automatically classified according to their distances in two categories, such as clustering algorithms.

Reinforcement Learning is the third method of machine learning. It is less popular compared with supervised and unsupervised methods. Under, Reinforcement learning, an agent learns to behave in an unknown scenario through the signals of reward and punishment provided by a critic. Different to supervised learning, the reward and punishment signals give less information, in most of the cases only failure or success. The final objective of the agent is to maximize the total reward obtained in a complete learning episode. Figure 1.3 illustrate the process of reinforcement learning.

Fig. 1.1 Schematic representation of the supervised learning

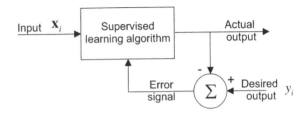

Fig. 1.2 Process of unsupervised learning. Data are automatically classified according to their distances in two categories, such as clustering algorithms

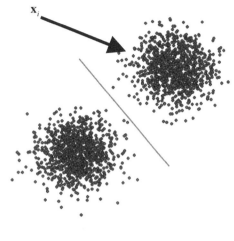

Fig. 1.3 Process of reinforcement learning

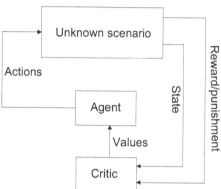

1.3 Classification

Classification considers the problem of determining categorical labels for unlabeled patterns based on observations. Let $(\mathbf{x}_1, y_1), \ldots, (\mathbf{x}_N, y_N)$ be observations of d-dimensional continuous patterns, i.e., $\mathbf{x}_i \in \mathbb{R}^d$ with discrete labels $y_1, \ldots, y_N$. The objective in classification is to obtain a functional model f that allows a reasonable prediction of unknown class labels y' for a new pattern $\mathbf{x}'$. Patterns without labels should be assigned to labels of patterns that are enough similar, e.g., that are close to the target pattern in data space, that come from the same distribution, or that lie on the same side of a separating decision function. But learning from observed patterns can be difficult. Training sets can be noisy, important features may be unknown, similarities between patterns may not be easy to define, and observations may not be sufficiently described by simple distributions. Further, learning functional models can be tedious task, as classes may not be linearly separable or may be difficult to separate with simple rules or mathematical equations.

1.3.1   Nearest Neighbors

The Nearest neighbor (NN) method is the most popular method used in machine learning for classification. Its best characteristic is its simplicity. It is based on the idea that the closest patterns to a target pattern x', for which we seek the label, deliver useful information of its description. Based on this idea, NN assigns the class label of the majority of the k-nearest patterns in data space. Figure 1.4 show the classification process under the NN method, considering a 4-nearest approach. Analyzing Fig. 1.4, it is clear that the novel pattern x' will be classified as element of the class A, since most of the nearest element are of the A category.

1.4   Parametric and Non-parametric Models

The objective of a machine learning algorithm is to obtain a functional model f that allows a reasonable prediction or description of a data set. There are many ways to define such models, but the most important distinction is this: does the model have a fixed number of parameters, or does the number of parameters grow with the amount of training data? The former is called a parametric model, and the latter is called a nonparametric model. Parametric models have the advantage of often being faster to use, but the disadvantage of making stronger assumptions about the nature of the data distributions. Nonparametric models are more flexible, but often computationally intractable for large datasets. We will give examples of both kinds of models in the sections below. We focus on supervised learning for simplicity, although much of our discussion also applies to unsupervised learning. Figure 1.5 represents graphically the architectures from both approaches.

Fig. 1.4 Classification process under the NN method, considering a 4-nearest approach

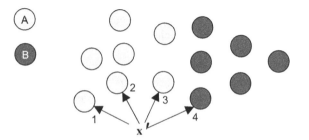

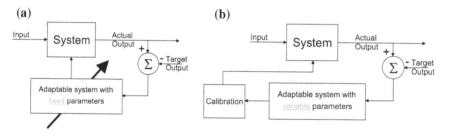

(a)   **(b)**

Fig. 1.5 Graphical representation of the learning process in Parametric and non-parametric models

1.5   Overfitting

The objective of learning is to obtain better predictions as outputs, being they class labels or continuous regression values. The process to know how successfully the algorithm has learnt is to compare the actual predictions with known target labels, which in fact is how the training is done in supervised learning. If we want to generalize the performance of the learning algorithm to examples that were not seen during the training process, we obviously can't test by using the same data set used in the learning stage. Therefore, it is necessary a different data, a test set, to prove the generalization ability of the learning method. This test set is used by the learning algorithm and compared with the predicted outputs produced during the learning process. In this test, the parameters obtained in the learning process are not modified.

In fact, during the learning process, there is at least as much danger in over-training as there is in under-training. The number of degrees of variability in most machine learning algorithms is huge—for a neural network there are lots of weights, and each of them can vary. This is undoubtedly more variation than there is in the function we are learning, so we need to be careful: if we train for too long, then we will overfit the data, which means that we have learnt about the noise and inaccuracies in the data as well as the actual function. Therefore, the model that we learn will be much too complicated, and won't be able to generalize.

Figure 1.6 illustrates this problem by plotting the predictions of some algorithm (as the curve) at two different points in the learning process. On the Fig. 1.6a the curve fits the overall trend of the data well (it has generalized to the underlying general function), but the training error would still not be that close to zero since it passes near, but not through, the training data. As the network continues to learn, it will eventually produce a much more complex model that has a lower training error (close to zero), meaning that it has memorized the training examples, including any noise component of them, so that is has overfitted the training data (see Fig. 1.6b).

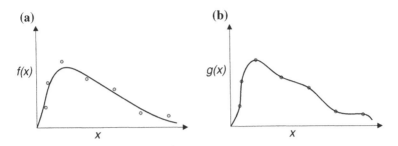

Fig. 1.6 Examples of **a** generalization and **b** overfitting

We want to stop the learning process before the algorithm overfits, which means that we need to know how well it is generalizing at each iteration. We can't use the training data for this, because we wouldn't detect overfitting, but we can't use the testing data either, because we're saving that for the final tests. So we need a third set of data to use for this purpose, which is called the validation set because we're using it to validate the learning so far. This is known as cross-validation in statistics. It is part of model selection: choosing the right parameters for the model so that it generalizes as well as possible.

1.6   The Curse of Dimensionality

The NN classifier is simple and can work quite well, when it is given a representative distance metric and an enough training data. In fact, it can be shown that the NN classifier can come within a factor of 2 of the best possible performance if $N \rightarrow \infty$.

However, the main problem with NN classifiers is that they do not work well with high dimensional data **x**. The poor performance in high dimensional settings is due to the curse of dimensionality.

To explain the curse, we give a simple example. Consider applying a NN classifier to data where the inputs are uniformly distributed in the d-dimensional unit cube. Suppose we estimate the density of class labels around a test point $\mathbf{x}'$ by "growing" a hyper-cube around $\mathbf{x}'$ until it contains a desired fraction F of the data points. The expected edge length of this cube will be $e_d(F) = F^{1/d}$. If $d = 10$ and we want to compute our estimate on 10 % of the data, we have $e_{10}(0.1) = 0.8$, so we need to extend the cube 80 % along each dimension around $\mathbf{x}'$. Even if we only use 1 % of the data, we find $e_{10}(0.01) = 0.63$, see Fig. 1.7. Since the entire range of the data is only 1 along each dimension, we see that the method is no longer very local, despite the name "nearest neighbor". The trouble with looking at neighbors that are so far away is that they may not be good predictors about the behavior of the input-output function at a given point.

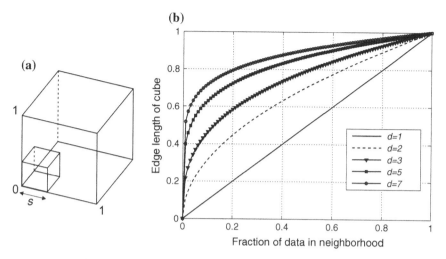

Fig. 1.7 Illustration of the curse of dimensionality. **a** We embed a small cube of side s inside a larger unit cube. **b** We plot the edge length of a cube needed to cover a given volume of the unit cube as a function of the number of dimensions

1.7 Bias-Variance Trade-Off

An inflexible model is defined as a mathematical formulation that involve few parameters. Due to the few parameters, they have several problems to fit data well. On the other hand, flexible models integrates several parameters with better modeling capacities [4]. The sensibility of inflexible models to variations during the learning process is relatively moderate in comparison to flexible models. Similarly, inflexible models have comparatively few modeling capacities, but are often easy to interpret. For example, linear models formulate linear relationships between their parameters, which are easy to describe with their coefficients, e.g.,

$$f(\mathbf{x}_i) = \alpha_0 + \alpha_1 x_i^1 + \cdots + \alpha_d x_i^d \qquad (1.1)$$

The coefficients α represents the adaptable parameters that formulate relationships easy to interpret. Optimizing the coefficients of a linear model is easier than fitting an arbitrary function with several adapting parameters.

Linear models do not suffer from overfitting as they are less depending on slight changes in the training set. They have low variance that is the amount by which the model changes, if using different training data sets. Such models have large errors when approximating a complex problem corresponding to a high bias. Bias is a measure for the inability of fitting the model to the training patterns. In contrast, flexible methods have high variance, i.e., they vary a lot when changing the training set, but have low bias, i.e., they better adapt to the observations.

Fig. 1.8 Illustration of
bias-variance compromise

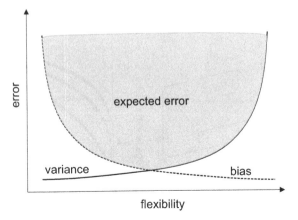

Figure 1.8 illustrates the bias-variance trade-off. On the x-axis the model complexity increases from left to right. While a method with low flexibility has a low variance, it usually suffers from high bias. The variance increases while the bias decreases with increasing model flexibility. The effect changes in the middle of the plot, where variance and bias cross. The expected error is minimal in the middle of the plot, where bias and variance reach a similar level. For practical problems and data sets, the bias-variance trade-off has to be considered when the decision for a particular method is made.

1.8   Data into Probabilities

Figure 1.9 shows the measurements of a feature $\mathbf{x}$ for two different classes, C_1 and C_2. Members of class C_2 tend to have larger values of feature $\mathbf{x}$ than members of class C_1, but there is some overlap between both classes. Under such conditions, the correct class is easy to predict at the extremes of the range of each class, but what to do in the middle where is unclear [3].

Fig. 1.9 Histograms of
feature $\mathbf{x}$ values against their
probability $p(\mathbf{x})$ for two
classes

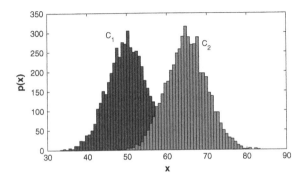

Assuming that we are trying to classify the writing letters 'a' and 'b' based on their height (as it is shown in Fig. 1.10). Most of the people write the letter 'a' smaller than their 'b', but not everybody. However, in this example, other class of information can be used to solve this classification problem. We know that in English texts, the letter 'a' is much more common than the letter 'b'. If we see a letter that is either an 'a' or a 'b' in normal writing, then there is a 75 % chance that it is an 'a.' We are using prior knowledge to estimate the probability that the letter is an 'a': in this example, $p(C_1) = 0.75$, $p(C_2) = 0.25$. If we weren't allowed to see the letter at all, and just had to classify it, then if we picked 'a' every time, we'd be right 75 % of the time.

In order to give a prediction, it is necessary to know the value $\mathbf{x}$ of the discriminant feature. It would be a mistake to use only the occurrence (a priori) probabilities $p(C_1)$ and $p(C_2)$. Normally, a classification problem is formulated through the definition of a data set which contains a set of values of $\mathbf{x}$ and the class of each exemplar. Under such conditions, it is easy to calculate the value of $p(C_1)$ (we just count how many times out of the total the class was C_1 and divide by the total number of examples), and also another useful measurement: the *conditional probability* of C_1 given that $\mathbf{x}$ has value X: $p(C_1|X)$. The conditional probability tells us how likely it is that the class is C_1 given that the value of $\mathbf{x}$ is X. So in Fig. 1.9 the value of $p(C_1|X)$ will be much larger for small values of X than for large values. Clearly, this is exactly what we want to calculate in order to perform classification. The question is how to get to this conditional probability, since we can't read it directly from the histogram. The first thing that we need to do to get these values is to quantize the measurement $\mathbf{x}$, which just means that we put it into one of a discrete set of values $\{X\}$, such as the bins in a histogram. This is exactly what is plotted in Fig. 1.8. Now, if we have lots of examples of the two classes, and the histogram bins that their measurements fall into, we can compute $p(C_i, X_j)$, which is the joint probability, and tells us how often a measurement of C_i fell into histogram bin X_j. We do this by looking in histogram bin X_j, counting the number of elements of C_i, and dividing by the total number of examples of any class.

We can also define $p(X_j|C_i)$, which is a different conditional probability, and tells us how often (in the training set) there is a measurement of X_j given that the example is a member of class C_i. Again, we can just get this information from the

Fig. 1.10 Letters "a" and "b" in the pixel context

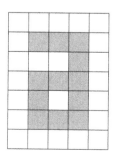 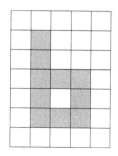

histogram by counting the number of examples of class C_i in histogram bin X_j and dividing by the number of examples of that class there are (in any bin).

So we have now worked out two things from our training data: the joint probability $p(C_i, X_j)$ and the conditional probability $p(X_j|C_i)$. Since we actually want to compute $p(C_i|X_j)$ we need to know how to link these things together. As some of you may already know, the answer is Bayes' rule, which is what we are now going to derive. There is a link between the joint probability and the conditional probability. It is:

$$p(C_i|X_j) = p(X_j|C_i) \cdot p(C_i) \tag{1.2}$$

or equivalently:

$$p(C_i, X_j) = p(C_i|X_j) \cdot p(X_j) \tag{1.3}$$

Clearly, the right-hand side of these two equations must be equal to each other, since they are both equal to $p(C_i, X_j)$, and so with one division we can write:

$$p(C_i|X_j) = \frac{p(X_j|C_i)p(C_i)}{p(X_j)} \tag{1.4}$$

This is Bayes' rule. If you don't already know it, learn it: it is the most important equation in machine learning. It relates the *posterior probability* $p(C_i|X_j)$ with the *prior* probability $p(C_1)$ and *class-conditional* probability $p(X_j|C_i)$ The denominator (the term on the bottom of the fraction) acts to normalize everything, so that all the probabilities sum to 1. It might not be clear how to compute this term. However, if we notice that any observation X_k has to belong to some class C_i, then we can marginalize over the classes to compute:

$$p(X_k) = \sum_i p(X_k|C_i) \cdot P(C_i) \tag{1.5}$$

The reason why Bayes' rule is so important is that it lets us obtain the posterior probability—which is what we actually want—by calculating things that are much easier to compute. We can estimate the prior probabilities by looking at how often each class appears in our training set, and we can get the class-conditional probabilities from the histogram of the values of the feature for the training set. We can use the posterior probability to assign each new observation to one of the classes by picking the class C_i where:

$$p(C_i|\mathbf{x}) > p(C_j|\mathbf{x}) \ \forall i \neq j \tag{1.6}$$

where $\mathbf{x}$ is a vector of feature values instead of just one feature. This is known as the *maximum* a posteriori or *MAP* hypothesis, and it gives us a way to choose which class to choose as the output one. The question is whether this is the right thing to do. There has been quite a lot of research in both the statistical and machine learning literatures into what is the right question to ask about our data to perform classification, but we are going to skate over it very lightly.

The MAP question is; what is the most likely class given the training data? Suppose that there are three possible output classes, and for a particular input the posterior probabilities of the classes are $p(C_1|\mathbf{x}) = 0.35$, $p(C_2|\mathbf{x}) = 0.45$, $p(C_3|\mathbf{x}) = 0.2$. The MAP hypothesis therefore tells us that this input is in class C_2, because that is the class with the highest posterior probability. Now suppose that, based on the class that the data is in, we want to do something. If the class is C_1 or C_3 then we do action 1, and if the class is C_2 then we do action 2. As an example, suppose that the inputs are the results of a blood test, the three classes are different possible diseases, and the output is whether or not to treat with a particular antibiotic. The MAP method has told us that the output is C_2.

As an example, suppose that the inputs are the results of a blood test, the three classes are different possible diseases, and the output is whether or not to treat with a particular antibiotic. The MAP method has told us that the output is C_2, and so we will not treat the disease. But what is the probability that it does not belong to class C_2, and so should have been treated with the antibiotic? It is $1 - p(C_2) = 0.55$. So the MAP prediction seems to be wrong: we should treat with antibiotic, because overall it is more likely. This method where we take into account the final outcomes of all of the classes is called the *Bayes' Optimal Classification*. It minimizes the probability of misclassification, rather than maximizing the posterior probability.

References

1. Kramer, O.: Machine Learning for Evolution Strategy. Springer, Switzerland (2016)
2. Kevin, P.: Murphy, Machine Learning: A Probabilistic Perspective. MIT Press, London (2011)
3. Marsland, S.: Machine Learning, An Algorithm Perspective. CRC Press, Boca Raton (2015)
4. Mola, A., Vishwanathan, A.: Introduction to Machine Learning. Cambridge University Press, Cambridge (2008)

Chapter 2
Optimization

2.1 Definition of an Optimization Problem

The vast majority of image processing and pattern recognition algorithms use some form of optimization, as they intend to find some solution which is "best" according to some criterion. From a general perspective, an optimization problem is a situation that requires to decide for a choice from a set of possible alternatives in order to reach a predefined/required benefit at minimal costs [1].

Consider a public transportation system of a city, for example. Here the system has to find the "best" route to a destination location. In order to rate alternative solutions and eventually find out which solution is "best," a suitable criterion has to be applied. A reasonable criterion could be the distance of the routes. We then would expect the optimization algorithm to select the route of shortest distance as a solution. Observe, however, that other criteria are possible, which might lead to different "optimal" solutions, e.g., number of transfers, ticket price or the time it takes to travel the route leading to the fastest route as a solution.

Mathematically speaking, optimization can be described as follows: Given a function $f : S \rightarrow \mathbb{R}$ which is called the objective function, find the argument which minimizes f:

$$x^* = \arg \min_{x \in S} f(x) \qquad (2.1)$$

S defines the so-called solution set, which is the set of all possible solutions for the optimization problem. Sometimes, the unknown(s) x are referred to design variables. The function f describes the optimization criterion, i.e., enables us to calculate a quantity which indicates the "quality" of a particular x.

In our example, S is composed by the subway trajectories and bus lines, etc., stored in the database of the system, x is the route the system has to find, and the optimization criterion $f(x)$ (which measures the quality of a possible solution) could

© Springer International Publishing AG 2017
D. Oliva and E. Cuevas, *Advances and Applications of Optimised Algorithms in Image Processing*, Intelligent Systems Reference Library 117, DOI 10.1007/978-3-319-48550-8_2

calculate the ticket price or distance to the destination (or a combination of both), depending on our preferences.

Sometimes there also exist one or more additional constraints which the solution x^* has to satisfy. In that case we talk about constrained optimization (opposed to unconstrained optimization if no such constraint exists). As a summary, an optimization problem has the following components:

- One or more design variables x for which a solution has to be found
- An objective function $f(x)$ describing the optimization criterion
- A solution set S specifying the set of possible solutions x
- (optional) One or more constraints on x

In order to be of practical use, an optimization algorithm has to find a solution in a reasonable amount of time with reasonable accuracy. Apart from the performance of the algorithm employed, this also depends on the problem at hand itself. If we can hope for a numerical solution, we say that the problem is well-posed. For assessing whether an optimization problem is well-posed, the following conditions must be fulfilled:

1. A solution exists.
2. There is only one solution to the problem, i.e., the solution is unique.
3. The relationship between the solution and the initial conditions is such that small perturbations of the initial conditions result in only small variations of x^*.

2.2 Classical Optimization

Once a task has been transformed into an objective function minimization problem, the next step is to choose an appropriate optimizer. Optimization algorithms can be divided in two groups: derivative-based and derivative-free [2].

In general, $f(x)$ may have a nonlinear form respect to the adjustable parameter x. Due to the complexity of $f(\cdot)$, in classical methods, it is often used an iterative algorithm to explore the input space effectively. In iterative descent methods, the next point x_{k+1} is determined by a step down from the current point x_k in a direction vector $\mathbf{d}$:

$$x_{k+1} = x_k + \alpha \mathbf{d}, \tag{2.2}$$

where α is a positive step size regulating to what extent to proceed in that direction. When the direction d in Eq. 2.1 is determined on the basis of the gradient ($\mathbf{g}$) of the objective function $f(\cdot)$, such methods are known as gradient-based techniques.

The method of steepest descent is one of the oldest techniques for optimizing a given function. This technique represents the basis for many derivative-based methods. Under such a method, Eq. 2.3 becomes the well-known gradient formula:

$$x_{k+1} = x_k - \alpha \mathbf{g}(f(x)), \qquad (2.3)$$

However, classical derivative-based optimization can be effective as long the objective function fulfills two requirements:

- The objective function must be two-times differentiable.
- The objective function must be uni-modal, i.e., have a single minimum.

A simple example of a differentiable and uni-modal objective function is

$$f(x_1, x_2) = 10 - e^{-\left(x_1^2 + 3 \cdot x_2^2\right)} \qquad (2.4)$$

Figure 2.1 shows the function defined in Eq. 2.4.

Unfortunately, under such circumstances, classical methods are only applicable for a few types of optimization problems. For combinatorial optimization, there is no dentition of differentiation.

Furthermore, there are many reasons why an objective function might not be differentiable. For example, the "floor" operation in Eq. 2.5 quantizes the function in Eq. 2.4, transforming Fig. 2.1 into the stepped shape seen in Fig. 2.2. At each step's edge, the objective function is non-differentiable:

$$f(x_1, x_2) = \text{floor}\left(10 - e^{-\left(x_1^2 + 3 \cdot x_2^2\right)}\right) \qquad (2.5)$$

Even in differentiable objective functions, gradient-based methods might not work. Let us consider the minimization of the Griewank function as an example.

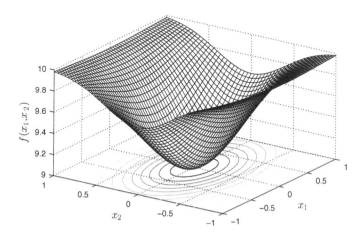

Fig. 2.1 Uni-modal objective function

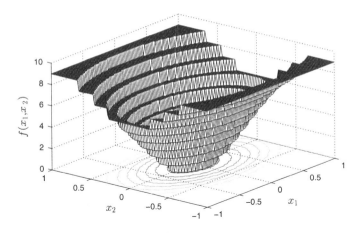

Fig. 2.2 A non-differentiable, quantized, uni-modal function

$$\begin{aligned} \text{minimize} \quad & f(x_1,x_2) = \tfrac{x_1^2+x_2^2}{4000} - \cos(x_1)\cos\!\left(\tfrac{x_2}{\sqrt{2}}\right) + 1 \\ \text{subject to} \quad & -30 \le x_1 \le 30 \\ & -30 \le x_2 \le 30 \end{aligned} \tag{2.6}$$

From the optimization problem formulated in Eq. 2.6, it is quite easy to understand that the global optimal solution is $x_1 = x_2 = 0$. Figure 2.3 visualizes the function defined in Eq. 2.6. According to Fig. 1.3, the objective function has many local optimal solutions (multimodal) so that the gradient methods with a randomly generated initial solution will converge to one of them with a large probability.

Considering the limitations of gradient-based methods, image processing and pattern recognition problems make difficult their integration with classical optimization methods. Instead, some other techniques which do not make assumptions and which can be applied to wide range of problems are required [3].

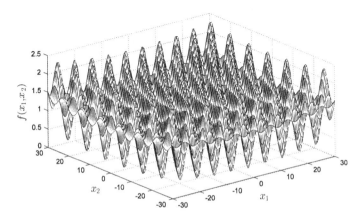

Fig. 2.3 The Griewank multi-modal function

2.3 Evolutionary Computation Methods

Evolutionary computation (EC) [4] methods are derivative-free procedures, which do not require that the objective function must be neither two-times differentiable nor uni-modal. Therefore, EC methods as global optimization algorithms can deal with non-convex, nonlinear, and multimodal problems subject to linear or nonlinear constraints with continuous or discrete decision variables.

The field of EC has a rich history. With the development of computational devices and demands of industrial processes, the necessity to solve some optimization problems arose despite the fact that there was not sufficient prior knowledge (hypotheses) on the optimization problem for the application of an classical method. In fact, in the majority of image processing and pattern recognition cases, the problems are highly nonlinear, or characterized by a noisy fitness, or without an explicit analytical expression as the objective function might be the result of an experimental or simulation process. In this context, the EC methods have been proposed as optimization alternatives.

A EC technique is a general method for solving optimization problems. It uses an objective function in an abstract and efficient manner, typically without utilizing deeper insights into its mathematical properties. EC methods do not require hypotheses on the optimization problem nor any kind of prior knowledge on the objective function. The treatment of objective functions as "black boxes" [5] is the most prominent and attractive feature of EC methods.

EC methods obtain knowledge about the structure of an optimization problem by utilizing information obtained from the possible solutions (i.e., candidate solutions) evaluated in the past. This knowledge is used to construct new candidate solutions which are likely to have a better quality.

Recently, several EC methods have been proposed with interesting results. Such approaches uses as inspiration our scientific understanding of biological, natural or social systems, which at some level of abstraction can be represented as optimization processes [6]. These methods include the social behavior of bird flocking and fish schooling such as the Particle Swarm Optimization (PSO) algorithm [7], the cooperative behavior of bee colonies such as the Artificial Bee Colony (ABC) technique [8], the improvisation process that occurs when a musician searches for a better state of harmony such as the Harmony Search (HS) [9], the emulation of the bat behavior such as the Bat Algorithm (BA) method [10], the mating behavior of firefly insects such as the Firefly (FF) method [11], the social-spider behavior such as the Social Spider Optimization (SSO) [12], the simulation of the animal behavior in a group such as the Collective Animal Behavior [13], the emulation of immunological systems as the clonal selection algorithm (CSA) [14], the simulation of the electromagnetism phenomenon as the electromagnetism-Like algorithm [15], and the emulation of the differential and conventional evolution in species such as the Differential Evolution (DE) [16] and Genetic Algorithms (GA) [17], respectively.

2.3.1 Structure of an Evolutionary Computation Algorithm

From a conventional point of view, an EC method is an algorithm that simulates at some level of abstraction a biological, natural or social system. To be more specific, a standard EC algorithm includes:

1. One or more populations of candidate solutions are considered.
2. These populations change dynamically due to the production of new solutions.
3. A fitness function reflects the ability of a solution to survive and reproduce.
4. Several operators are employed in order to explore an exploit appropriately the space of solutions.

The EC methodology suggest that, on average, candidate solutions improve their fitness over generations (i.e., their capability of solving the optimization problem). A simulation of the evolution process based on a set of candidate solutions whose fitness is properly correlated to the objective function to optimize will, on average, lead to an improvement of their fitness and thus steer the simulated population towards the global solution.

Most of the optimization methods have been designed to solve the problem of finding a global solution of a nonlinear optimization problem with box constraints in the following form:

$$\begin{aligned} \text{maximize} \quad & f(\mathbf{x}), \quad \mathbf{x} = (x_1, \dots, x_d) \in \mathbb{R}^d \\ \text{subject to} \quad & \mathbf{x} \in \mathbf{X} \end{aligned} \qquad (2.7)$$

where $f : \mathbb{R}^d \to \mathbb{R}$ is a nonlinear function whereas $\mathbf{X} = \{\mathbf{x} \in \mathbb{R}^d | l_i \leq x_i \leq u_i, i = 1, \dots, d.\}$ is a bounded feasible search space, constrained by the lower (l_i) and upper (u_i) limits.

In order to solve the problem formulated in Eq. 2.6, in an evolutionary computation method, a population $\mathbf{p}^k (\{\mathbf{p}_1^k, \mathbf{p}_2^k, \dots, \mathbf{p}_N^k\})$ of N candidate solutions (individuals) evolves from the initial point $(k = 0)$ to a total *gen* number iterations $(k = gen)$. In its initial point, the algorithm begins by initializing the set of N candidate solutions with values that are randomly and uniformly distributed between the pre-specified lower (l_i) and upper (u_i) limits. In each iteration, a set of evolutionary operators are applied over the population $\mathbf{p}^k$ to build the new population $\mathbf{p}^{k+1}$. Each candidate solution $\mathbf{p}_i^k$ $(i \in [1, \dots, N])$ represents a d-dimensional vector $\{p_{i,1}^k, p_{i,2}^k, \dots, p_{i,d}^k\}$ where each dimension corresponds to a decision variable of the optimization problem at hand. The quality of each candidate solution $\mathbf{p}_i^k$ is evaluated by using an objective function $f(\mathbf{p}_i^k)$ whose final result represents the fitness value of $\mathbf{p}_i^k$. During the evolution process, the best candidate solution $\mathbf{g}$ $(g_1, g_2, \dots g_d)$ seen so-far is preserved considering that it represents the best available solution. Figure 2.4 presents a graphical representation of a basic cycle of a EC method.

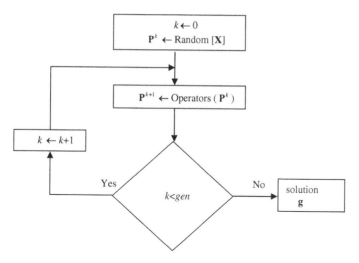

Fig. 2.4 The basic cycle of a EC method

2.4 Optimization and Image Processing and Pattern Recognition

Image processing and pattern recognition emerge as a part of decision support systems in industrial, medical, military applications, etc. Several methods have been proposed in the literature to solve image processing and pattern recognition tasks. They require some method-specific parameters to be tuned optimally to achieve the best performance, and this necessity makes the methods candidate to be converted into optimization problems.

Optimization has a fundamental importance in solving many problems in image processing and pattern recognition. Such a fact is evident from a quick look at the special issues, congresses and specialized journals in these areas, where significant percentage of the papers uses optimization techniques.

Classical optimization methods often face great difficulties while dealing with images or systems containing noise and distortions. Under such conditions, the use of evolutionary computation approaches has been recently extended to address challenging real-world image processing and pattern recognition problems [18].

Image processing and pattern recognition are both, dynamic and fast moving fields of research. On the other hand, each new approach developed by engineers, mathematicians and computer scientists is quickly identified, understood and as-simulated in order to be applied in image processing and pattern recognition. In this book we strive to bring some state of the art techniques using recent results in evolutionary computation, applied to challenging and significant problems in image processing and pattern recognition [19].

Evolutionary computation methods are vast and have many variants. Consequently, there exist a rich amount of literature in this subject, including textbooks, tutorials, and journal papers that cover in detail practically every aspect of the field. The great amount of information available makes it difficult for no specialist to explore the literature and find the right optimization technique for a specific image or patter recognition application. This fact becomes evident that any attempt to present the whole area of evolutionary computation in detail would be a daunting task, probably doomed to failure. This task would be even more difficult if the goal is to understand the applications of evolutionary methods in the context of image processing and pattern recognition. For this reason only the Electromagnetism-Like algorithm has been selected to be considered in this book.

References

1. Akay, B., Karaboga, D.: A survey on the applications of artificial bee colony in signal, image, and video processing. SIViP **9**(4), 967–990 (2015)
2. Yang, X.-S.: Engineering Optimization. Wiley, New York (2010)
3. Treiber, M.A.: Optimization for Computer Vision An Introduction to Core Concepts and Methods. Springer, Berlin (2013)
4. Simon, D.: Evolutionary Optimization Algorithms. Wiley, New York (2013)
5. Blum, C., Roli, A.: Metaheuristics in combinatorial optimization: overview and conceptual comparison. ACM Comput. Surv. (CSUR) **35**(3), 268–308 (2003). doi:10.1145/937503. 937505
6. Nanda, S.J., Panda, G.: A survey on nature inspired metaheuristic algorithms for partitional clustering, Swarm Evol. Comput. **16**, 1–18 (2014)
7. Kennedy, J., Eberhart, R.: Particle swarm optimization. In: Proceedings of the 1995 IEEE International Conference on Neural Networks, vol. 4, pp. 1942–1948, Dec 1995
8. Karaboga, D.: An Idea based on honey bee swarm for numerical optimization. Technical Report-TR06. Engineering Faculty, Computer Engineering Department, Erciyes University (2005)
9. Geem, Z.W., Kim, J.H., Loganathan, G.V.: A new heuristic optimization algorithm: harmony search. Simulations **76**, 60–68 (2001)
10. Yang, X.S.: A new metaheuristic bat-inspired algorithm. In: Cruz, C., González, J., Krasnogor, G.T.N., Pelta, D.A. (eds.) Nature Inspired Cooperative Strategies for Optimization (NISCO 2010), Studies in Computational Intelligence, vol. 284, pp. 65–74. Springer, Berlin (2010)
11. Yang, X.S.: Firefly algorithms for multimodal optimization. In: Stochastic Algorithms: Foundations and Applications, SAGA 2009, Lecture Notes in Computer Sciences, vol. 5792, pp. 169–178 (2009)
12. Cuevas, E., Cienfuegos, M., Zaldívar, D., Pérez-Cisneros, M.: A swarm optimization algorithm inspired in the behavior of the social-spider. Expert Syst. Appl. **40**(16), 6374–6384 (2013)
13. Cuevas, E., González, M., Zaldivar, D., Pérez-Cisneros, M., García, G.: An algorithm for global optimization inspired by collective animal behaviour. Discrete Dyn. Nat. Soc. 638275 (2012)
14. de Castro, L.N., von Zuben, F.J.: Learning and optimization using the clonal selection principle. IEEE Trans. Evol. Comput. **6**(3), 239–251 (2002)
15. Birbil, Ş.I., Fang, S.C.: An electromagnetism-like mechanism for global optimization. J. Glob. Optim. **25**(1), 263–282 (2003)
16. Storn, R., Price, K.: Differential evolution—a simple and efficient adaptive scheme for global optimisation over continuous spaces. Technical Report TR-95–012, ICSI, Berkeley, CA

17. Goldberg, D.E.: Genetic Algorithm in Search Optimization and Machine Learning. Addison-Wesley, (1989)
18. Cuevas, E., Zaldivar, D., Pérez-Cisneros, M., Ramírez-Ortegón, M.: Circle detection using discrete differential evolution optimization. Pattern Anal. Appl. **14**(1), 93–107 (2011)
19. Oliva, D., Cuevas, E., Pajares, G., Zaldivar, D., Osuna, V.: A multilevel thresholding algorithm using electromagnetism optimization. Neurocomputing **139**, 357–381 (2014)

Chapter 3
Electromagnetism—Like Optimization Algorithm: An Introduction

3.1 Introduction to Electromagnetism-Like Optimization (EMO)

In recent years the study of Evolutionary Computation Algorithms (ECA) has attracted the attention of researchers from different areas. Different to the traditional implementations they have been developed some alternatives that are inspired not only in biological processes but also in natural mechanisms. For example, the methods based on physical phenomena. One of those techniques is the Electromagnetism-Like Optimization algorithm (EMO) [1]. The analogy of EMO is the superposition principle and the Coulomb law for charged particles. One of the main advantages on EMO is that its operators are easy to implement and they maintain a good balance between exploration and exploitation in the search space. In this chapter the features of EMO will be analyzed. However, it is important to note that although initially EMO may have certain similarities with other meta-heuristics techniques, it has several features that improve its performance. EMO is a method that does not require gradient operations, neither crossover or mutation operations and its application does not imply the use of special numerical system. Moreover, with a reduced population it is possible to find the global optimal in few iterations.

The EMO algorithm is a relatively new population based method. EMO was proposed by Birbil and Fang [1] to solve continuous optimization models with bounded variables. In other words, EMO can solve optimization problems with bounded search spaces. The algorithm mimics the attraction-repulsion mechanism between charged particles in a magnetic field. Under the EMO context, each particle represents a solution and possess an amount of charge that is proportional to the solution quality (defined by the objective function). On the same way, the candidate solutions are defined by position vectors that assign the real positions of the particles in a multidimensional space. Moreover the values of the objective

© Springer International Publishing AG 2017
D. Oliva and E. Cuevas, *Advances and Applications of Optimised Algorithms in Image Processing*, Intelligent Systems Reference Library 117,
DOI 10.1007/978-3-319-48550-8_3

function of each particle are computed considering such position vectors. Each particle exerts attraction or repulsion forces over the other population members. The resultant force that acts over an element of the population is used to update its next position. Clearly, the basic idea of EMO is to move the particles to an optimal solution using the attraction and repulsion forces according with the electromagnetism principles.

Different to other similar ECA as: Genetic Algorithms (GA) [2], Particle Swarm Optimization (PSO) [3], Differential Evolution (DE) [4] ot Ant Colony Optimization [5] where their population members share information or materials between them. EMO assumes that each particle is influenced by the rest of the population similar to the PSO or ACO operators. Although EMO shares some features with PSO and ACO there exist recent works that exhibit the good EMO convergence maintaining the accuracy to find the optimal solutions [1, 6]. Moreover, the EMO convergence has been successfully proved solving several engineering problems. Standout among other the following implementations:

- Flowshop scheduling problems [7].
- Communications [8].
- Vehicle routing problems [9].
- Electronic circuit design [10].
- Neural network training [11].
- Image processing [12, 13].
- Medical image [14].

Considering the implementations of EMO in the state-of-the-art, it is possible to say that EMO is a promising optimization technique, with a huge field of applications. In the following subsection, first the basic structure of EMO will be analyzed. After that their implementation using a numerical example will be presented.

3.2 Optimization Inspired in Electromagnetism

As was mention previously, the EMO algorithm solve optimization problems defined as:

$$\text{minimize} \quad f(\mathbf{X}), \quad \mathbf{X} \in [l, u] \tag{3.1}$$

where l and u are the lower and upper bounds of the search space $\Re^n$ and it is defined as $[l, u] = \left\{ \mathbf{X} \in \Re^d | l_k \leq x_k \leq u_k, k = 1, 2, \ldots, d \right\}$. d is the dimension of the $\mathbf{X}$ vector that is candidate solution, finally $f()$ is the function to optimize. It is possible to say that the described above are the features of the optimization problem. Considering these facts, EMO employs two basic processes for the optimization: the first one performs a random exploration of the search space, meanwhile the second process exploit the selected local points. For this purpose

EMO uses the operators based on the electromagnetism theory. The balanced use of both exploration and exploitation processes guarantee that the algorithm converges in the optimal values of the objective functions that are highly attractive. Based on these key processes, the EMO algorithms has four phases to achieve the global optimization [1]. Each stage is described as follows:

- **Initialization**: a set of m particles are randomly taken from a feasible space $\Re^d$, defined by the upper (u) and lower (l) limits (Eq. 3.1).
- **Local search**: it is performed a search for the best value in a defined neighbourhood of a particle $\mathbf{X}^p$, where $p \in (1, 2, \ldots, m)$ and m corresponds to the total number of elements in the population.
- **Total force vector**: the charges and forces for each element of the population is computed based on their objective function value.
- **Movement**: each particle is displaced according to the total force.

Each step of EMO will be described in the following subsections they include pseudocodes for a better comprehension.

3.2.1   Initialization

In this stage a set $\mathbf{X}$ of m candidate solutions is randomly generated, the elements of this set is are considered initial solutions. Each member of the population is considered as candidate solution or under the EMO context a charged particle. It is assumed that all particles in the population are uniformly distributed in the search space. The best particle or solution is the one with the best value of the objective function. The initialization procedure ends when all the m particles are initialized and evaluated in the objective function. According with the EMO authors the initialization process is defined in the Algorithm 3.1.

Algorithm 3.1 EMO Initialization	
1.	for $i = 1$ to m do
2.	for $k = 1$ to d do
3.	$\lambda \leftarrow rand\,(0,1)$
4.	$x_k^i \leftarrow l_k + \lambda\,(u_k - l_k)$
5.	end for
6.	Compute $f\left(\mathbf{X}^i\right)$
7.	end for

3.2.2 Local Search

This stage tries to improve the solutions previously found, in some cases this phase could be modified. Considering this it is possible to formulate a classification of the EMO based algorithms: EMO without local search; (a) EMO with local search only for the best particle and (b) EMO with local search over all the particles. The last one is the more used in the EMO implementations.

The local search works considering a determined number of evaluations of the objective function *LSITER* and a parameter for the search neighbourhood (δ). This procedure is very simply and it works as follows: a point $\mathbf{x}^p \in \mathbf{X}$ is assigned to a temporal variable $\mathbf{y}$. The temporal variable is used to store the main value of the population. After that a selected coordinate d modified using a step length that is computed using a random number (λ_1) and δ. The element stored in $\mathbf{y}$ is displaced in the direction of the step length the sign of this direction is selected randomly using λ_2. If the obtained value after evaluate $\mathbf{y}$ in the objective function is better and the number of *LSITER* is achieved, the point $\mathbf{x}^p$ is updated with $\mathbf{y}$ otherwise $\mathbf{x}^p$ holds its value. Finally the best value is updated in the population, the pseudocode of this method is described in the Algorithm 3.2

Algorithm 3.2: Local Search for EMO			
1.	$count \leftarrow 1$	12.	$y_d = y_d - \lambda_2 \cdot (Length)$
2.	$Length \leftarrow \delta \big(\max (u_k - l_k) \big)$	13.	**end if**
3.	**for** $i = 1$ **to** m **do**	14.	**if** $f(\mathbf{y}) < f(\mathbf{x}^i)$ **then**
4.	**for** $k = 1$ **to** d **do**	15.	$\mathbf{x}^p \leftarrow \mathbf{y}$
5.	$\lambda_1 \leftarrow \mathrm{rand}(0,1)$	16.	$count \leftarrow LSITER - 1$
6.	**while** $count < LSITER$ **do**	17.	**end if**
7.	$\mathbf{y} \leftarrow \mathbf{x}^i$	18.	$count \leftarrow count + 1$
8.	$\lambda_2 \leftarrow \mathrm{rand}(0,1)$	19.	**end while**
9.	**if** $\lambda_1 > 0.5$ **then**	20.	**end for**
10.	$y_d = y_d + \lambda_2 \cdot (Length)$	21.	**end for**
11.	**else**	22.	$\mathbf{x}^{best} \leftarrow \arg\min \big(f(\mathbf{x}^i), \mathbf{x}^i \in \mathbf{X} \big)$

In general, the local search applied over all the particles can reduce the risk to fall in a local optima. The disadvantage is that this process increases the computational time, an efficient alternative to this problem is the use of local search only in the best particle. This modification helps to preserve the efficiency and the computational accuracy. However, it reduces the exploitation of the search space. In the local search the step length represents an important factor that depends on the limits of each dimension and determines the performance of the search.

3.2.3   Total Force Vector

The total force vectors computation is based on the superposition principle of the electromagnetism principle. The superposition principle establishes that "the force exerted on a particle by other particle is inversely proportional to the distance between the points and directly proportional to their charges product" [15]. An example of the superposition principle is presented in Eq. 3.2 to compute the total force of a charge from the system presented in Fig. 3.1.

$$\overrightarrow{F_{i,3}} = \left(\frac{q_3 \cdot q_i}{4\pi\varepsilon_0\varepsilon_r r^2}\right) \cdot \overrightarrow{e_r}, \quad i = 1, 2 \tag{3.2}$$

Under the EMO approach each particle is moved according with the Coulomb law. This law employs the force produced between the particles that depends on the charge that each one possess. Such charge is determined by its objective function value and it is obtained as follows:

$$q^p = \exp\left(-n\frac{f(\mathbf{X}^p) - f(\mathbf{X}^{best})}{\sum\limits_{h=1}^{m}\left(f(\mathbf{X}^h) - f(\mathbf{X}^{best})\right)}\right), \forall p \tag{3.3}$$

where n denotes the dimension of $\mathbf{X}^p$ and m represents the population size, meanwhile the vector $\mathbf{X}^h$ corresponds to the hth particle of the population. $\mathbf{X}^p$ is the current element which its charge q is computed. exp is the exponential function and $f(\cdot)$ is the objective function. It is important to mention that a high dimensional optimization problem will requires a bigger population. In the Eq. 3.3 $\mathbf{X}^{best}$ is the best particle according with objective function and it has the higher charge.

The attraction force exerted by the best particle over other given particle, is inversely proportional to the distance between them. Therefore, this particle attracts other particles with worst objective function values and repels the ones with best values. The resultant force that exist between the particles, determines the

Fig. 3.1 Superposition principle

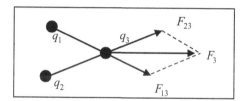

medication of the particles position in the optimization process. The force of each member of the EMO population is computed using the Coulomb law and the superposition principle described in Eq. 3.4.

$$
\mathbf{F}^p = \sum_{h \neq p}^{m}
\begin{cases}
(\mathbf{X}^h - \mathbf{X}^p) \frac{q^p q^h}{\|\mathbf{X}^h - \mathbf{X}^p\|^2} & \text{if } f(\mathbf{X}^h) < f(\mathbf{X}^p) \\
(\mathbf{X}^p - \mathbf{X}^h) \frac{q^p q^h}{\|\mathbf{X}^h - \mathbf{X}^p\|^2} & \text{if } f(\mathbf{X}^h) \geq f(\mathbf{X}^p)
\end{cases}, \forall p \qquad (3.4)
$$

where $f(\mathbf{X}^h) < f(\mathbf{X}^p)$ corresponds to the attraction and $f(\mathbf{X}^h) \geq f(\mathbf{X}^p)$ represents the repulsion of charges, an example is shown in Fig. 3.2. By definition the resultant force of each particle is proportional to the charges product and inversely proportional to the distance between the particles. To make this process mathematically coherent the Eq. 3.4 must be normalized as is explained in Eq. 3.5. Finally the Algorithm 3.3 presents the necessary steps to compute the total force vector.

$$
\overline{\mathbf{F}^p} = \frac{\mathbf{F}^p}{\|\mathbf{F}^p\|}, \quad \forall p \qquad (3.5)
$$

	Algorithm 3.3: Total force vector
1.	**for** $i = 1$ **to** m **do**
2.	$q^i \leftarrow \exp\left(-n \dfrac{f(\mathbf{x}^i) - f(\mathbf{x}^{best})}{\sum_{h=1}^{m}\left(f(\mathbf{x}^h) - f(\mathbf{x}^{best})\right)}\right)$
3.	$\mathbf{F}^i \leftarrow 0$
4.	**end for**
5.	**for** $i = 1$ **to** m **do**
6.	**for** $k = 1$ **to** d **do**
7.	**if** $f(\mathbf{X}^j) < f(\mathbf{X}^i)$ **then**
8.	$\mathbf{F}^i \leftarrow \mathbf{F}^i + (\mathbf{x}^j - \mathbf{x}^i)\dfrac{q^i q^j}{\|\mathbf{x}^j - \mathbf{x}^i\|^2}$, (Atraction)
9.	**else**
10.	$\mathbf{F}^i \leftarrow \mathbf{F}^i - (\mathbf{x}^j - \mathbf{x}^i)\dfrac{q^i q^j}{\|\mathbf{x}^j - \mathbf{x}^i\|^2}$, (Repulsion)
11.	**end if**
12.	**end for**
13.	**end for**

Fig. 3.2 Coulomb law

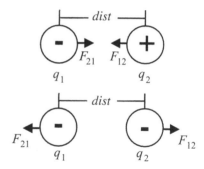

3.2.4 Movement

Once the total force is computed, the next step is to find the new positions of the population members according with the principle of attraction-repulsion of charges. In this way, considering the resultant force of each particle the new position is computed by the following equation:

$$\mathbf{x}^p = \begin{cases} \mathbf{x}^p + \lambda \cdot \mathbf{F}^p \cdot \left(u_d - \mathbf{x}_d^p \right) & \text{if} \quad \mathbf{F}^p > 0 \\ \mathbf{x}^p + \lambda \cdot \mathbf{F}^p \cdot \left(\mathbf{x}_d^p - l_d \right) & \text{if} \quad \mathbf{F}^p \leq 0 \end{cases}, \quad \forall p \neq best \qquad (3.6)$$

where λ is a random value uniformly distributed between [0,1], on the other hand u_d and l_d are the upper and lower limits of the d dimension respectively. If the force is determined as repulsion the particle $\mathbf{x}^p$ is moved to the upper limit, otherwise the movement is to the lower limit. The best particle is not affected by the movement because it has the absolute attraction and repels or attract the remainder elements of the population.

The iterative EMO process ends when a stop criteria is achieved. For this purpose there are commonly used two single rules: (1) a predefined number of iteration is achieved, or (2) when a value of $f\left(\mathbf{x}^{best}\right)$ is optimal in one sense. The local search, total force vector and movement represent the exploitation process of EMO. The algorithm 3.4 shows each step of the movement procedure.

Algorithm 3.4: Movement	
1.	**for** $i = 1$ **to** m **do**
2.	**if** $i \neq mejor$ **then**
3.	$\lambda \leftarrow rand(0,1)$
4.	$\mathbf{F}^i = \dfrac{\mathbf{F}^i}{\left\| \mathbf{F}^i \right\|}$
5.	**for** $k = 1$ **to** d **do**
6.	**if** $\mathbf{F}_k^i > 0$ **then**
7.	$x_k^i \leftarrow x_k^i + \lambda \cdot F_k^i \cdot \left(u_k - x_k^i \right)$
8.	**else**
9.	$x_k^i \leftarrow x_k^i + \lambda \cdot F_k^i \cdot \left(x_k^i - l_k \right)$
10.	**end if**
11.	**end for**
12.	**end if**
13.	**end for**

3.3 A Numerical Example Using EMO

The efficacy of global optimization algorithms is commonly tested using mathematical function; even some sets of benchmark functions had been crated [16]. The Rosenbrock function is a typical two dimensional problem used to test the evolutionary computation algorithms. This function is defined as follows:

$$f(x,y) = (1 - x)^2 + 100\left(y - x^2\right)^2 \tag{3.7}$$

Since this is a benchmark function its parameter are already well know and they are described in Table 3.1.

Table 3.1 Limits and global minima of Rosenbrock function

Upper limit (u)	Lower limit (l)	Global minima
$x = 2$	$x = -2$	$x^{best} = 1$
$y = 2$	$y = -2$	$y^{best} = 1$

Table 3.2 EMO parameter used for the Rosenbrock function

Population size (N)	Length step (δ)	$Iter_{max}$
10	0.001	50

The values of Table 3.1 are used to define the search space used by EMO. The parameters of EMO are initialized considering the suggested values of the original paper [1] (see Table 3.2).

The iterative process of EMO can starts once its parameter are initialized. The first procedure is generate a random population of m particles. An example of a population for the Rosenbrock objective function is shown in Table 3.3.

The initial population values are evolved using all the EMO processes. When a stop criteria is satisfied the algorithm finish the iterations. For this example the stop criteria is considered when a maximum number of iterations is achieved. The final population and the Rosenbrock function are plotted in Fig. 3.3.

3.4 EMO Modifications

The aim of this book is to show that Electromagnetism-like algorithm is a good alternative that could be used to solve different problems from image processing. However EMO has an inconvenient, the Local Search procedure that make the algorithm computationally expensive.

In order to overcome this situation there are proposed different modification based on some approaches. Before present the most important modification it is important to mention that the EMO performances is subject to the No Free Lunch theorem [17]. Such theorem basically explains that an optimization algorithm cannot guarantee to have the best performance in all the applications without sacrifice the accuracy or vice versa. In this way not all the modifications work efficiently in all the problems and it is the same situation of EMO. For example a global optimization problem without constrictions could be solved with a single EMO algorithm, probably we can only apply the local search to the best value and obtain the global optima. On the other hand the same algorithm cannot found the optimal of a constrained optimization problem.

In the following chapters this book presents two modification of EMO for specific problems in image processing. Moreover this section explains the most important modifications selected from the state-of-the art.

Table 3.3 Possible EMO population generated by the initialization procedure

x	-1.6109	0.7127	-0.8361	1.7886	1.0436	-0.6682	0.4100	-1.7861	1.5771	-0.8316
y	-1.8580	0.8332	-1.9183	1.5444	-0.2486	-1.0087	-1.4834	0.6282	-0.8488	-0.8769
$f(x, y)$	1989.79	10.65	688.43	274.40	178.96	214.53	273.09	664.14	1113.24	249.36

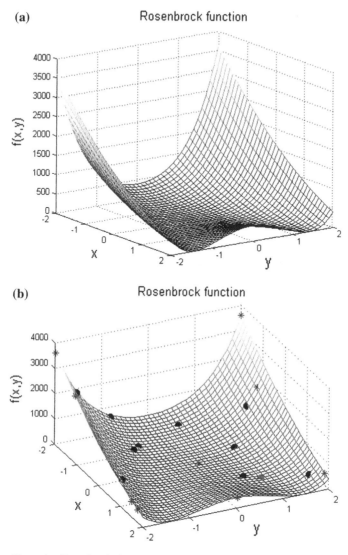

Fig. 3.3 a Show the Rosenbrock function, **b** the initial population (*black dots*), the evolved population (*blue cross*) and the global minima (*red dot*)

The algorithms selected for this section are:

- Hybridizing EMO with descent search (HEMO) [18];
- EMO with fixed search pattern (FEMO) [19];

3.4.1 Hybridizing EMO with Descent Search (HEMO)

The hybridizing EMO with descent search is a modified algorithm firstly proposed [17]. This algorithm integrates the feasibility and dominance (FAD) rules [20–22]. Such rules are an alternative to the penalty functions in gradient-based or derivative-free methods for constrained optimization problems. The integration of EMO and FAD rules is considered a simple task [23]. The FAD rules are used to select the best point in specific situations:

1. Among two feasible points, here the point that has the better objective function is selected.
2. Any feasible point is preferred to any infeasible solution.
3. Among two infeasible points, the one that has the smaller constraints violations in proffered.

Those rules can provide a good results in some optimization problems, but the convergence cannot be guarantee using them. In this sense stronger conditions must be proposed in order to select the best elements of the population. The challenge is consider and adequate reduction and constraints violations to have an efficient optimization algorithm.

FAD rules implemented on EMO algorithm

The FAD rules are implemented in the step where the charge of each element is computed. The standard EMO uses the entire population and their objective function value to compute the charges. The HEMO uses the rule presented in Eq. 3.8 to penalize the infeasible solutions.

$$\text{fitness}(x) = \begin{cases} f(x) & \text{if} \quad x \text{ is feasible} \\ f_{\text{max}} + \text{CV}(x) & \text{otherwise} \end{cases} \qquad (3.8)$$

where f_{max} is the maximum function value of the feasible particles on the current population. Meanwhile the function $\text{CV}(x) = \|\max(0, g(x))\|_2$ measures the constraints violations. The use of CV considering for example a point x with $\text{CV}(x) = 0$ is feasible. The particle is infeasible if $\text{CV}(x) > 0$, if the current population has no feasible points f_{max} is set to 0. Once the rules are applied the EMO algorithm continues computing the charges and total force vector with the procedures presented in the previous sections.

Reduction conditions in HEMO

The modified HEMO considers a sufficient reduction in either the constraints violation or in the objective function value. There are used to detect the best particle in the population to attract the rest to promising regions and to guarantee that the population is moving to the best point. The penalty function in deterministic approaches are used to ensure the progress the solution by enforcing a sufficient reduction in the penalty function. However such methods requires the use of a merit function or update the penalty parameter. To avoid this situation there is introduced a filter technique [24] that helps to achieve the global optima from any arbitrary approximation (Eq. 3.9). In the filter methodology a particle is selected if there exist enough reduction conditions.

$$CV(y) \leq (1 - \gamma)CV(x) \quad \text{or} \quad f(y) \leq f(x) - \gamma CV(x) \quad \text{for} \quad \gamma \in (0, 1) \quad (3.9)$$

The selection of the sufficient reduction conditions is implemented as follows. If two particles x^i and x^j are feasible, x^i is selected only if a sufficient decrease (or increase) in f is verified using Eq. 3.10.

$$f(x^i) \leq (1 - \gamma)f(x^j) \quad (3.10)$$

$\gamma \in (0, 1)$ is a fixed constant. If only one particle is feasible or the two compared particles are infeasible then x^i is selected if the rule of Eq. 3.11 is satisfied.

$$CV(x^i) \leq (1 - \gamma)CV(x^j) \quad (3.11)$$

In practical terms the constraint violations are firstly measured for all the population. If the point is feasible the objective function is evaluated. Otherwise is value is not required.

In the standard version of EMO the movement uses a RNG function which value is computed using the limits of the search space. Basically the RNG is a step length used to move the particles according the total force vector. The HEMO uses a single rule presented in Eq. 3.12 to compute the value of RNG depending if the total force vector has a value that can help to find the best solution.

$$RNG^i = \min_{1 \leq k \leq n} \left(RNG_k^i \right) \equiv \begin{cases} \frac{(u_k - x_k^i)}{F_k^i} & \text{if} \quad F_k^i > 0 \\ \frac{(l_k - x_k^i)}{F_k^i} & \text{if} \quad F_k^i < 0 \\ M & \text{if} \quad F_k^i = 0 \end{cases} \quad (3.12)$$

where M is a large positive number to penalize a total force vector that not helps in the optimization process.

Local descent search

One of the main advantages of the HEMO is the integration of a new local search algorithm. This mechanism is a derivative-free heuristic method that produces an approximate descent direction and aims to generate a new solution around the best particle of the population. The local descent search (LDS) needs to explore the neighborhood of the best element of the population using Eq. 3.13.

$$x_k^{rand,i} = x_k^{best} \pm \lambda \varepsilon_r, \quad \text{for} \quad k = 1, 2, \ldots, n \tag{3.13}$$

where $\lambda \sim U(0, 1)$, ε_r is a sufficiently small positive value and $i = 1,2$. Once the space around the best particle is explored it is necessary to obtain the descent direction dir for the fitness function (Eq. 3.14). This process is performed considering the two points computed by Eq. 3.13.

$$dir = -\frac{1}{\sum\limits_{j=1}^{2} |\Delta_j|} \sum\limits_{i=1}^{2} \Delta_i \frac{x^{best} - x^{rand,i}}{\|x^{best} - x^{rand,i}\|} \tag{3.14}$$

In Eq. 3.14 $\Delta_j = fitness(x^{best}) - fitness(x^{rand,j})$ the theory about this direction function is available in [25]. The use of Eq. 3.14 can produce a descent direction, however it is restricted to one of the following conditions:

1. The points compared are randomly generated in a small neighborhood around the best element of the population.
2. The points are in equal distance to the best element of the population.

A trial point then is generated along the descent direction with a prescribed step size using Eq. 3.15.

$$y = x^{best} + s \cdot RNG \cdot dir \tag{3.15}$$

where $s \in (0, 1]$ represents the step size and RNG is computed according Eq. 3.12— The selection of the step size employs a backtracking strategy. Algorithm 3.5 details the entire procedure of the Local descent search.

	Algorithm 3.5 Local descent search
1.	$flag \leftarrow 1$, $s \leftarrow 1$, $iteration \leftarrow 0$
2.	**while** $iteration \leq LSITER$ **do**
3.	**if** $flag = 1$ **then**
4.	Create two random points around x^{best} (Eq. 3.13)
5.	Obtain the descent direction using Eq. 3.14
6.	**end if**
7.	Compute the trial point y (Eq. 3.15)
8.	**if** y and x^{best} are feasible **then**
9.	**if** $f(y) \leq (1-\gamma) f(x^{best})$ **then**
10.	$x^{best} \leftarrow y$, $s \leftarrow 1$, $flag \leftarrow 1$
11.	**else**
12.	$s \leftarrow s/2$, $flag \leftarrow 0$
13.	**end if**
14.	**else**
15.	**if** $CV(y) \leq (1-\gamma) CV(x^{best})$ **then**
16.	$x^{best} \leftarrow y$, $s \leftarrow 1$, $flag \leftarrow 1$
17.	**else**
18.	$s \leftarrow s/2$, $flag \leftarrow 0$
19.	**end if**
20.	**end if**
21.	$iteration \leftarrow iteration + 1$
22.	**end while**

3.4.2 EMO with Fixed Search Pattern (FEMO)

Another important modification of EMO is the FEMO that includes a local search based on the pattern search method of Hooke and Jeeves [19]. FEMO also employs a shrinking strategy that aims to reduce the population size along the iterative process.

The idea behind FEMO is that at the beginning the optimization problem requires a larger population. Meanwhile when the evolutionary process is in an advanced stage the population is near to the global solution. In this case a less number of elements are necessary to explore the search space. A reduction of the population size at different stages does not affect the convergence of the EMO.

Hooke and Jeeves pattern search method

The modification of the local search in FEMO is based on the assumption than in EMO, the local search procedure can be applied only to the best element of the population. This fact produces a direct search method with good capabilities of convergence. On the other hand, the pattern search (PS) is a popular method that can replace the classical local search procedure in EMO [1], [26–28]. PS has been extensively used in unconstrained and constrained optimization problems [28] for that reason it is simply to apply it to EMO.

The PS generates feasible points, the feasibility is verified using the following penalty function:

$$P(x) \equiv \begin{cases} f(x) & \text{if} \quad x \in \Omega \\ \infty & \text{otherwise} \end{cases} \tag{3.16}$$

In Eq. 3.16, any no feasible point is penalized using a very high or low value (depending on the optimization problem). That means that only the feasible points will be evaluated in the objective function. The PS method uses two different movements: the exploratory and the pattern move. In the exploratory move it is performed a coordinate search around the best population element using a step length δ. If for the new point y: $P(y) < P(x^{best})$, the current iteration is successful. Otherwise the value of δ is reduced. When a iteration is successful the vector $y - x^{best}$ is used as promising direction for the PS. In other words a new particle is generated using $y + (y - x^{best})$. If the new point is feasible the population is updated, otherwise the value is modified only with y. It is important to mention that a minimum step length value should be defined for the PS methodology.

Population shrinking strategy

The use of the shrinking strategy reduces the particles in the population as the optimization process is evolving. In this sense the number of evaluation function will be also reduced without affecting the accuracy of the final results. The challenge is to decide how the population is reduced. A strategy to solve this challenge is use a value that measures the concentration of points to compute the spread of the function value with respect to best value. The spread (SPR) used in FEMO is computed using Eq. 3.17.

$$SPR = \left(\frac{\sum\limits_{i=1}^{m} \left(f(x^i) - f\left(x^{best}\right) \right)^2}{m} \right)^{1/2} \tag{3.17}$$

The SPR is used to decide when is necessary to shrink the particles population. Basically the idea is shrink the population at a specific number of iterations. To decide when is possible to shrink it is used a SPR reference value $\left(SPR^{ref}\right)$, if the computed SPR is below of certain percentage of SPR^{ref} the population is shrink. Equation 3.18 present a rule used to define when the population is reduced.

$$SPR < \varepsilon SPR^{ref}, \quad \varepsilon = 0.1 \tag{3.18}$$

It is important to mention that this shrinking methodology is used only when the population has at least $2n$ particles considering n the number of dimension of the problem. The pseudocode of the shirking strategy included in the FEMO is presented in Algorithm 3.6

	Algorithm 3.6 FEMO
1.	Initialization()
2.	$iteration \leftarrow 1$
3.	$SPR^{ref} \leftarrow Compute(SPR)$
4.	**while** stop criteria is not satisfied **do**
5.	$F \leftarrow Calc\_Force(\ )$
6.	$Move(F)$
7.	$Local(LSITER, \delta)$
8.	**if** $m > 2n$ **then**
9.	$Compute(SPR)$
10.	**if** $SPR < \varepsilon SPR^{ref}$ **then**
11.	$m = m/2$ and discard m points
12.	$SPR^{ref} \leftarrow SPR$
13.	**end if**
14.	**end if**
15.	$iteration \leftarrow iteration + 1$
16.	**end while**

References

1. Birbil, Ş.I., Fang, S.C.: An electromagnetism-like mechanism for global optimization. J. Glob. Optim. **25**(1), 263–282 (2003)
2. De Jong, K.: Learning with genetic algorithms: an overview. Mach. Learn. **3**, 121–138 (1988)
3. Kennedy, J., Eberhart, R.: Particle swarm optimization. In: Proceedings of the IEEE International Conference on Neural Networks, 1995, vol. 4, pp. 1942–1948 (1995)
4. Storn, R., Price, K.: Differential evolution—a simple and efficient heuristic for global optimization over continuous spaces. J. Glob. Optim. **11**(4), 341–359
5. Dorigo, M., Maniezzo, V., Colorni, A.: The ant systems: optimization by a colony of cooperative agents. IEEE Trans. Man, Mach. Cybern. B **26**(1) (1996)
6. Birbil, Ş.I., Fang, S.C., Sheu, R.L.: On the convergence of a population-based global optimization algorithm. J. Glob. Optim. **30**(2–3), 301–318 (2004)
7. Naderi, B., Tavakkoli-Moghaddam, R., Khalili, M.: Electromagnetism-like mechanism and simulated annealing algorithms for flowshop scheduling problems minimizing the total weighted tardiness and makespan. Knowledge-Based Syst. **23**(2), 77–85 (2010)
8. Hung, H.L., Huang, Y.F.: Peak to average power ratio reduction of multicarrier transmission systems using electromagnetism-like method. Int. J. Innov. Comput. Inf. Control **7**(5), 2037–2050 (2011)
9. Yurtkuran, A., Emel, E.: A new hybrid electromagnetism-like algorithm for capacitated vehicle routing problems. Expert Syst. Appl. **37**(4), 3427–3433 (2010)
10. Jhang, J.Y., Lee, K.C.: Array pattern optimization using electromagnetism-like algorithm. AEU—Int. J. Electron. Commun. **63**, 491–496 (2009)
11. Lee, C.H., Chang, F.K.: Fractional-order PID controller optimization via improved electromagnetism-like algorithm. Expert Syst. Appl. **37**(12), 8871–8878 (2010)
12. Oliva, D., Cuevas, E., Pajares, G., Zaldivar, D.: Template matching using an improved electromagnetism-like algorithm. Appl. Intell. **41**, 791–807 (2014)
13. Cuevas, E., Oliva, D., Zaldivar, D., Pérez-Cisneros, M., Sossa, H.: Circle detection using electro-magnetism optimization. Inf. Sci. (Ny) **182**(1), 40–55 (2012)
14. Cuevas, E., Oliva, D., Díaz, M., Zaldivar, D., Pérez-Cisneros, M., Pajares, G.: White blood cell segmentation by circle detection using electromagnetism-like optimization. Comput. Math. Methods Med. **2013** (2013)
15. Cowan, E.W.: Basic Electromagnetism. Academic Press, New York (1968)
16. Dixon, G.P., Szego, L.C.: The global optimization problem: an introduction. In: Dixon, G.P., Szego, L.C. (eds.) Towards Global Optimization 2, pp. 1–15. North-Holland Publishing Company, Amsterdam (1978)
17. Wolpert, D.H., Macready, W.G.: No free lunch theorems for optimization. IEEE Trans. Evol. Comput. **1**(1), 67–82 (1997)
18. Rocha, A.M.A.C., Fernandes, E.M.G.P.: Hybridizing the electromagnetism-like algorithm with descent search for solving engineering design problems. Int. J. Comput. Math. **86**, 1932–1946 (2009)
19. Rocha, A.M.A.C., Fernandes, E.M.G.P.: A new electromagnetism-like algorithm with a population shrinking strategy. In: Proceedings of the 9th WSEAS International Conference on Mathematical and Computational methods in Science and Engineering, vol. 1, no. 3, pp. 45–50 (2007)
20. Deb, K.: An efficient constraint handling method for genetic algorithms. Comput. Methods Appl. Mech. Eng. **186**(2–4), 311–338 (2000)
21. Karaboga, D., Basturk, B.: Artificial bee colony (ABC) optimization algorithm for solving constrained optimization. Lnai **4529**, 789–798 (2007)
22. Zavala, A.E.M., Aguirre, A.H., Villa Diharce, E.R.: Particle evolutionary swarm optimization algorithm (PESO). In: Proceedings of the Mexican International Conference on Computer Science, vol. 2005, pp. 282–289 (2005)

23. Rocha, A.M.A.C., Fernandes, E.M.G.P.: Feasibility and dominance rules in the electromagnetism-like algorithm for constrained global optimization. Lecture Notes in Computer Science (including Subseries Lecture Notes in Artificial Intelligence Lecture Notes Bioinformatics), vol. 5073 LNCS, no. PART 2, pp. 768–783, 2008

24. Fletcher, R., Leyffer, S.: Nonlinear programming without a penalty function. Math. Program. **91**(2): 239–2369 (2002)

25. Hedar, A.-R., Fukushima, M.: Heuristic pattern search and its hybridization with simulated annealing for nonlinear global optimization. Optim. Methods Softw. **19**(3–4), 291–308 (2004)

26. Audet, C., Dennis, J.E.: Analysis of generalized pattern searches. SIAM J. Optim. **13**(3), 889–903 (2002)

27. Kolda, T., Lewis, R., Torczon, V.: Optimization by direct search: new perspectives on some classical and modern methods. SIAM Rev. **45**(3), 385–482 (2003)

28. Lewis, R., Torczon, V.: Pattern search algorithms for bound constrained minimization. SIAM J. Optim. **9**(4), 1082–1099 (1999)

Chapter 4
Digital Image Segmentation as an Optimization Problem

4.1 Introduction

Image processing has several applications in areas as medicine, industry, agriculture, etc. Most of all the methods of image processing require a first step called segmentation. This task consists in classify the image pixels depending on its gray (or RGB in each component) level intensity (histogram). In this way, several techniques had been studied [1, 2]. Thresholding is the easiest method for segmentation, it works taking a threshold (th) value and the pixels which intensity value is higher than th are labelled as the first class and the rest of the pixels correspond to the second class. When the image is segmented in two classes it is called bi-level thresholding (BT) and it is necessary only one th value. On the other hand, when pixels are separated in more than two classes it is called multilevel thresholding (MT) and there are required more than one th values [2, 3]. Threshold based methods are divided in parametric and nonparametric [3–5]. For parametric approaches it is necessary to estimate some parameters of a probability density function which models each class. Such approaches are time consuming and computationally expensive. On the other hand, the nonparametric employs a given criteria (between-class variance, entropy and error rate [6–8]) that must be optimized to determine the optimal threshold values. These approaches result an attractive option due their robustness and accuracy [9].

For bi-level thresholding there exist two classical methods, the first maximizes between classes variance and was proposed by Otsu [6]. The second proposed by Kapur [7] uses the maximization of the entropy to measures the homogeneity of the classes. Their efficiency and accuracy have been already proved for two segmentation classes [10]. Although both Otsu's and Kapur's can be expanded for multilevel thresholding, their computational complexity increases exponentially with each new threshold [10].

© Springer International Publishing AG 2017
D. Oliva and E. Cuevas, *Advances and Applications of Optimised Algorithms in Image Processing*, Intelligent Systems Reference Library 117, DOI 10.1007/978-3-319-48550-8_4

On the other hand, Tsallis entropy was proposed by Tsallis [11], it is called non-extensive entropy and could be considered as an extension of Shannon's entropy. There exist some studies that propose the existence of similarities between the entropies function of: Tsallis, Shannon's and Boltzmann/Gibbs [12–16]. They have some similarities due the use of entropic indexes. On the other hand, Portes de Albuquerque [13] proposes the use of TE for image segmentation. Similar to other non-parametric methods, TE requires a *th* that maximizes the entropy value to find this value is used the full search that is complex and highly time consuming.

As an alternative to classical methods, the MT problem has also been handled through evolutionary optimization methods. In general, they have demonstrated to deliver better results than those based on the classical techniques in terms of accuracy, speed and robustness. Numerous evolutionary approaches have been reported in the literature. Hammouche et al. [3] provides a survey of different evolutionary algorithms such as (Differential Evolution (DE), Simulated Annealing (SA), Tabu Search (TS) etc.), used to solve the Kaptur's and Otsu's problems. In [3, 17, 18], Genetic Algorithms-based approaches are employed to segment multi-classes. Similarly in [1, 6], Particle Swarm Optimization (PSO) [19] has been proposed for MT proposes, maximizing the Otsu's function. Other examples such [20–22] include Artificial Bee Colony (ABC) or Bacterial Foraging Algorithm (BFA) for image segmentation. Meanwhile, in [23] is presented a TH algorithm based on the classical Differential Evolution (DE) [24] and TE. Another interesting approach [25] applies the Particle Swarm Optimization algorithm (PSO) [19] to maximize the Tsallis entropy. The Artificial Bee Colony (ABC) [26] is used in [27] for multilevel thresholding with TE. New state-of-the-art evolutionary algorithms as Cuckoo Search (CSA) [26, 28] and Bacterial Foraging (BFA) [29] are recently proposed in [30, 31] for image TH using as objective function the Tsallis entropy.

This chapter introduces a multilevel threshold method based on the Electromagnetism-like Algorithm (EMO). EMO is a global optimization algorithm that mimics the electromagnetism law of physics. It is a population-based method which has an attraction-repulsion mechanism to evolve the members of the population guided by their objective function values [32]. The main idea of EMO is to move a particle through the space following the force exerted by the rest of the population. The force is calculated using the charge of each particle based on its objective function value. Unlike other meta-heuristics such as GA, DE, ABC and Artificial Immune System (AIS), where the population members exchange materials or information between each other, in EMO similar to heuristics such as PSO and Ant Colony Optimization (ACO) each particle is influenced by all other particles within its population. Although the EMO algorithm shares some characteristics to PSO and ACO, recent works have exhibited its better accuracy regarding optimal parameters [33–36], yet showing convergence [37]. In recent works, EMO has been used to solve different sorts of engineering problems such as flow-shop scheduling [38], communications [39], vehicle routing [40], array pattern optimization in circuits [41], neural network training [42], image processing [43] and control systems [44].

In this chapter, a segmentation method called Multilevel Threshold based on the EMO algorithm (MTEMO) is introduced. The algorithm takes random samples from a feasible search space which depends on the image histogram. Such samples build each particle in the EMO context. The quality of each particle is evaluated considering the objective function employed by the Otsu's, Kapur's or Tsallis method. Guided by this objective value the set of candidate solutions are evolved using the attraction-repulsion operators. The approach generates a multilevel segmentation algorithm which can effectively identify the threshold values of a digital image in a reduced number of iterations and decreasing the computational complexity of the original proposals. Experimental results show performance evidence of the implementation of EMO for digital image segmentation.

The rest of the chapter is organized as follows. Section 4.2 gives a simple description of the Otsu's, Kapur's and Tsallis methods. Section 4.3 explains the implementation of EMO for multilevel thresholding. Section 4.4 discusses experimental results and comparisons after test the MTEMO in a set benchmark images. Finally, the work is concluded in Sect. 4.5.

4.2   Image Multilevel Thresholding (MT)

Thresholding is a process in which the pixels of a gray scale image are divided in sets or classes depending on their intensity level (L). For this classification it is necessary to select a threshold value (th) and follows the simple rule of Eq. (4.1).

$$\begin{aligned} C_1 &\leftarrow p \quad \text{if} \quad 0 \leq p < th \\ C_2 &\leftarrow p \quad \text{if} \quad th \leq p < L-1 \end{aligned} \tag{4.1}$$

where p is one of the mxn pixels of the gray scale image I_g that can be represented in L gray scale levels $L = \{0, 1, 2, \ldots, L-1\}$. C_1 and C_2 are the classes in which the pixel p can be located, finally th is the threshold. The rule in Eq. (4.1) corresponds to bi-level thresholding and can be easily extended for multiple sets:

$$\begin{aligned} C_1 &\leftarrow p \quad \text{if} \quad 0 \leq p < th_1 \\ C_2 &\leftarrow p \quad \text{if} \quad th_1 \leq p < th_2 \\ C_i &\leftarrow p \quad \text{if} \quad th_i \leq p < th_{i+1} \\ C_n &\leftarrow p \quad \text{if} \quad th_n \leq p < L-1 \end{aligned} \tag{4.2}$$

where $\{ th_1 \quad th_2 \quad \ldots \quad th_i \quad th_{i+1} \quad th_k \}$ represent the different thresholds. The problem for both bi-level and multilevel thresholding is to select the th values that correctly identify the classes. Otsu's, Kapur's and Tsallis methods are well-known approaches for determining such values. Both methods propose a different objective function which must be maximized in order to find the optimal threshold values.

4.2.1   Between—Class Variance (Otsu's Method)

This is a nonparametric technique for thresholding proposed by Otsu [6] that employs the maximum variance value of the different classes as criterion to segment the image. Taking the L intensity levels from an intensity image or from each component of a RGB (red, green, blue) image, the probability distribution of the intensity values is computed as follows:

$$Ph_i^c = \frac{h_i^c}{NP}, \quad \sum_{i=1}^{NP} Ph_i^c = 1, \quad c = \begin{cases} 1,2,3 & \text{if} \quad \text{RGB Image} \\ 1 & \text{if} \quad \text{Gray scale Image} \end{cases} \tag{4.3}$$

where i is a specific intensity level ($0 \leq i \leq L - 1$), c is the component of the image which depends if the image is intensity or RGB whereas NP is the total number of pixels in the image. h_i^c (histogram) is the number of pixels that corresponds to the i intensity level in c. The histogram is normalized in a probability distribution Ph_i^c. For the simplest segmentation (bi-level) two classes are defined as:

$$C_1 = \frac{Ph_1^c}{\omega_0^c(th)}, \ldots, \frac{Ph_{th}^c}{\omega_0^c(th)} \quad \text{and} \quad C_2 = \frac{Ph_{th+1}^c}{\omega_1^c(th)}, \ldots, \frac{Ph_L^c}{\omega_1^c(th)} \tag{4.4}$$

where $\omega_0(th)$ and $\omega_1(th)$ are probabilities distributions for C_1 and C_2, Eq. (4.5).

$$\omega_0^c(th) = \sum_{i=1}^{th} Ph_i^c, \quad \omega_1^c(th) = \sum_{i=th+1}^{L} Ph_i^c \tag{4.5}$$

It is necessary to compute the mean levels μ_0^c and μ_1^c that define the classes using Eq. (4.6). Once those values are obtained the Otsu based between—class $\sigma_B^{2^c}$ is calculated using Eq. (4.7).

$$\mu_0^c = \sum_{i=1}^{th} \frac{iPh_i^c}{\omega_0^c(th)}, \quad \mu_1^c = \sum_{i=th+1}^{L} \frac{iPh_i^c}{\omega_1^c(th)} \tag{4.6}$$

$$\sigma_B^{2^c} = \sigma_1^c + \sigma_2^c \tag{4.7}$$

Notice that for both Eqs. (4.6) and (4.7), c depends on the type of image. Moreover σ_1^c and σ_2^c in Eq. (4.7) are the variances of C_1 and C_2 which are defined as:

$$\sigma_1^c = \omega_0^c \left(\mu_0^c + \mu_T^c \right)^2, \quad \sigma_2^c = \omega_1^c \left(\mu_1^c + \mu_T^c \right)^2 \tag{4.8}$$

where $\mu_T^c = \omega_0^c \mu_0^c + \omega_1^c \mu_1^c$ and $\omega_0^c + \omega_1^c = 1$. Based on the values σ_1^c and σ_2^c, Eq. (4.9) presents the objective function. Therefore, the optimization problem is reduced to find the intensity level that maximizes Eq. (4.9).

$$J(th) = \max(\sigma_B^{2c}(th)), \quad 0 \leq th \leq L - 1 \tag{4.9}$$

Otsu's method is applied for a single component of an image, which means for RGB images it is necessary to separate them in single component images. The previous description of this bi-level method can be extended for the identification of multiple thresholds. Considering k thresholds it is possible to separate the original image into k classes using Eq. (4.2), then it is necessary to compute the k variances and their respective elements. The objective function $J(th)$ in Eq. (4.9) can be rewritten for multiple thresholds as follows:

$$J(\mathbf{TH}) = \max(\sigma_B^{2c}(\mathbf{TH})), \quad 0 \leq th_i \leq L - 1, \quad i = 1, 2, \dots, k \tag{4.10}$$

where $\mathbf{TH} = [th_1, th_2, \dots, th_{k-1}]$, is a vector that contains multiple thresholds and the variances are computed using Eq. (4.11).

$$\sigma_B^{2c} = \sum_{i=1}^{k} \sigma_i^c = \sum_{i=1}^{k} \omega_i^c \left(\mu_i^c - \mu_T^c\right)^2 \tag{4.11}$$

Here i represents and specific class. ω_i^c and μ_j^c are respectively the probability of occurrence and the mean of a class, respectively. For MT such values are obtained as:

$$\omega_0^c(th) = \sum_{i=1}^{th_1} Ph_i^c$$

$$\omega_1^c(th) = \sum_{i=th_1+1}^{th_2} Ph_i^c$$

$$\vdots \qquad \vdots$$

$$\omega_{k-1}^c(th) = \sum_{i=th_k+1}^{L} Ph_i^c \tag{4.12}$$

and for the mean values:

$$\mu_0^c = \sum_{i=1}^{th_1} \frac{iph_i^c}{\omega_0^c(th_1)}$$

$$\mu_1^c = \sum_{i=th_1+1}^{th_2} \frac{iph_i^c}{\omega_0^c(th_2)}$$

$$\vdots \qquad \vdots$$

$$\mu_{k-1}^c = \sum_{i=th_k+1}^{L} \frac{iph_i^c}{\omega_1^c(th_k)} \tag{4.13}$$

Similar to the bi-level case, for MT using the Otsu's method c corresponds to the image components, RGB $c = 1, 2, 3$ and intensity $c = 1$.

4.2.2   *Entropy Criterion Method (Kapur's Method)*

Another nonparametric method used to determine the optimal threshold values was proposed by Kapur [7]. It is based on the entropy and the probability distribution of an image histogram, the purpose of this method is to find the optimal *th* that maximizes the overall entropy. The entropy of an image measures the compactness and separability among classes. In this sense when the optimal *th* value separates the classes properly the entropy has the maximum value. For the bi-level example the objective function of the Kapur's problem can be defined as:

$$J(th) = H_1^c + H_2^c, \quad c = \begin{cases} 1, 2, 3 & \text{if} \quad \text{RGB Image} \\ 1 & \text{if} \quad \text{Gray scale Image} \end{cases} \quad (4.14)$$

where the entropies H_1 and H_2 are computed using the following model:

$$H_1^c = \sum_{i=1}^{th} \frac{Ph_i^c}{\omega_0^c} \ln\left(\frac{Ph_i^c}{\omega_0^c}\right), \quad H_2^c = \sum_{i=th+1}^{L} \frac{Ph_i^c}{\omega_1^c} \ln\left(\frac{Ph_i^c}{\omega_1^c}\right) \quad (4.15)$$

Ph_i^c is the probability distribution of the intensity levels which is obtained using Eq. (4.3). $\omega_0(th)$ and $\omega_1(th)$ are probabilities distributions for C_1 and C_2. $\ln(\cdot)$ corresponds to the natural logarithm. Similar to the Otsu's method the entropy-based approach can be extended for multiple threshold values, for such situation it is necessary to divide the image into k classes using the similar number of thresholds. Under such conditions, the new objective function is defined as:

$$J(\mathbf{TH}) = \sum_{i=1}^{k} H_i^c, \quad c = \begin{cases} 1, 2, 3 & \text{if} \quad \text{RGB Image} \\ 1 & \text{if} \quad \text{Gray scale Image} \end{cases} \quad (4.16)$$

where $\mathbf{TH} = [th_1, th_2, \ldots, th_{k-1}]$, is a vector that contains the multiple thresholds. Each entropy is computed separately with its respective *th* value, Eq. (4.17) is expanded for k entropies.

$$H_1^c = \sum_{i=1}^{th_1} \frac{Ph_i^c}{\omega_0^c} \ln\left(\frac{Ph_i^c}{\omega_0^c}\right),$$

$$H_2^c = \sum_{i=th_1+1}^{th_2} \frac{Ph_i^c}{\omega_1^c} \ln\left(\frac{Ph_i^c}{\omega_1^c}\right),$$

$$\vdots \qquad\qquad \vdots \qquad\qquad\qquad (4.17)$$

$$H_k^c = \sum_{i=th_k+1}^{L} \frac{Ph_i^c}{\omega_{k-1}^c} \ln\left(\frac{Ph_i^c}{\omega_{k-1}^c}\right)$$

Here the values of the probability occurrence $(\omega_0^c, \omega_1^c, \ldots, \omega_{k-1}^c)$ of the k classes are obtained using Eq. (4.12) and the probability distribution Ph_i^c with Eq. (4.3). Finally to separate the pixels in the respective classes it is necessary to use Eq. (4.2).

4.2.3 Tsallis Entropy

The entropy is defined in thermodynamic to measure the order of irreversibility in the universe. The concept of entropy physically expresses the amount of disorder of a system [13, 30]. In information theory Shannon redefine the theory proposed by Boltzmann/Gibbs and employ the entropy to measure the uncertainty regarding information of a system [13]. In other words, is possible the quantitatively measurement of the amount of information produced by a process.

In basics of entropy in a discrete system is taking a probability distribution defined as $p = \{p_i\}$ that represents the probability of find the system in a possible state i. Notice that $0 \leq p_i \leq 1$ and $\sum_{i=1}^{k} p_i = 1, k$ is the total number of states. In addition, a physical or information system can be decomposed in two statistical independent subsystems A and B with probabilities p^A and p^B then the probability of the composed system is $p^{A+B} = p^A \cdot p^B$. Such definition has been verified using the extensive property (additivity) (Eq. 4.18) proposed by Shannon [13, 30].

$$S(A+B) = S(A) + S(B) \qquad (4.18)$$

Tsallis proposed a generalized form of statistics based on the related concepts and the multi-fractal theory. The Tsallis entropic form is an important tool used to describe the thermostatistical properties of non-extensive systems and is defined as:

$$S_q = \frac{1 - \sum_{i=1}^{k} (p_i)^q}{q - 1} \qquad (4.19)$$

where S is the Tsallis entropy, $q \to 1$ is the Tsallis entropic index that represents the degree of nonextensivity and k is the total number of possibilities of the system. Since Tsallis entropy is nonextensive is necessary to redefine the additivity entropic rule of Eq. (4.18).

$$S_q(A+B) = S_q(A) + S_q(B) + (1 - q) \cdot S_q(A) \cdot S_q(B) \qquad (4.20)$$

Since image segmentation has non-additive information content, it is possible to use the Tsallis entropy to find the best thresholds [13]. A digital gray scale image has k gray levels that are defined by the histogram. The easiest thresholding considers to classes divided by a one *th* (bi-level), to solve this problem is considered the probability distribution of the gray levels ($p_i = p_1, p_2, \ldots p_k$). For each class A and B two probability distributions are created (Eq. 4.21) one for each class using th.

$$p_A = \frac{p_1}{p^A}, \frac{p_2}{p^A}, \ldots \frac{p_{th}}{p^A} \quad \text{and} \quad p_B = \frac{p_1}{p^B}, \frac{p_2}{p^B}, \ldots \frac{p_k}{p^B} \tag{4.21}$$

where:

$$P^A = \sum_{i=1}^{th} p_i \quad \text{and} \quad P^B = \sum_{i=th+1}^{k} p_i \tag{4.22}$$

The TE for class A and class B is defined as follows:

$$S_q^A(th) = \frac{1 - \sum_{i=1}^{th} \left(\frac{p_i}{p^A}\right)^q}{q - 1}, \quad S_q^B(th) = \frac{1 - \sum_{i=th+1}^{k} \left(\frac{p_i}{p^B}\right)^q}{q - 1} \tag{4.23}$$

TE value depends directly on the parameter th and it maximizes the information measured between two classes. If the value of $S_q(th)$ is maximized it means that the th is the optimal threshold value. In order to verify the efficiency of the selected th in Eq. (4.24) is proposed an objective function using Eq. (4.20).

$$TH_{opt}(th) = \arg \max \left(S_q^A(th) + S_q^B(th) + (1 - q) \cdot S_q^A(th) \cdot S_q^B(th) \right) \tag{4.24}$$

The previous description of this bi-level method can be extended for the identification of multiple thresholds. Considering nt thresholds it is possible separate the original image into nt classes. Under this approach the optimization problem turns into multidimensional situation. The solutions are conformed as $\mathbf{TH}_j = [th_1, th_2, \ldots th_{nt}]$. For each class is computed an entropy using the Tsallis methodology and the objective function is redefined as follows:

$$\begin{aligned} TH_{opt}(\mathbf{th}) = \arg \max \Big(S_q^1(th_1) + S_q^2(th_2) + \cdots \\ + S_q^{nt}(th_{nt}) + (1 - q) \cdot S_q^1(th_1) \cdot S_q^2(th_2) \cdots\cdots S_q^{nt}(th_{nt}) \Big) \end{aligned} \tag{4.25}$$

where:

$$\begin{aligned} S_q^1(th_1) = \frac{1 - \sum_{i=1}^{th_1} \left(\frac{p_i}{p^1}\right)^q}{q - 1}, \quad S_q^2(th) = \frac{1 - \sum_{i=th_1+1}^{th_2} \left(\frac{p_i}{p^2}\right)^q}{q - 1}, \ldots, \\ S_q^{nt}(th) = \frac{1 - \sum_{i=th_{nt-1}+1}^{th_{nt}} \left(\frac{p_i}{p^{nt}}\right)^q}{q - 1} \end{aligned} \tag{4.26}$$

specific class. However there exist an extra class it means that exist $nt + 1$ classes. The extra class is considered *default* class because it is computed from nt to k (Eq. 4.27).

$$S_q^{def}(th_k) = \frac{1 - \sum_{i=th_{nt}+1}^{k} \left(\frac{p_i}{P^k}\right)^q}{q - 1} \tag{4.27}$$

4.3   Multilevel Thresholding Using EMO (MTEMO)

This section presents the implementation of the EMO for multilevel thresholding in digital image. The optimization algorithm is used to optimize the objective functions previously presented. Firstly the general particle representation for MT is described. After that the implementation is divided in two: for Otsu's and Kapur's methods and for the Tsallis entropy.

4.3.1   Particle Representation

Each particle uses k different elements, as decision variables, in the optimization algorithm. Such decision variables represent a different threshold point th used for the segmentation. Therefore, the complete population is represented as:

$$\mathbf{S}_t = [\mathbf{TH}_1^c, \mathbf{TH}_2^c, \ldots, \mathbf{TH}_N^c], \quad \mathbf{TH}_i^c = \left[th_1^c, th_2^c, \ldots, th_k^c\right]^T \tag{4.28}$$

where t represents the iteration number, T refers to the transpose operator, N is the size of the population and $c = 1, 2, 3$ for RGB images or $c = 1$ for intensity images.

4.3.2   EMO Implementation with Otsu's and Kapur's Methods

The presented segmentation algorithm has been implemented considered the two different objective functions, Otsu and Kapur. Therefore, the EMO algorithm was coupled with the Otsu and Kapur functions, producing two different segmentation algorithms. The implementation of such algorithms can be summarized into the following steps:

Step 1: Read the image I and if it RGB separate it into I_R, I_G and I_B. If the I is gray scale store it into I_{Gr}. $c = 1, 2, 3$ for RGB images or $c = 1$ for gray scale images.

Step 2:   Obtain histograms: for RGB images h^R, h^G, h^B and for gray scale images h^{Gr}.

Step 3:   Calculate the probability distribution using Eq. (4.3) and the histograms.

Step 4:   Initialize the EMO parameters: $Iter_{max}$, $Iter_{local}$, δ, k and N.

Step 5:   Initialize a population $\mathbf{S}_t^c$ of N random particles with k dimensions.

Step 6:   Compute the values ω_i^c and μ_i^c. Evaluate $\mathbf{S}_t^c$ in the objective function $J(\mathbf{S}_t^c)$ Eqs. (4.10) or (4.16) depending on the thresholding method (Otsu or Kapur respectively).

Step 7:   Compute the charge of each particle and the total force vector according with the method described in Chap. 3.

Step 8:   Move the entire population $\mathbf{S}_t^c$ along the total force vector.

Step 9:   Apply the local search to the moved population and select the best elements of this search based on their objective function values.

Step 10: The t index is increased in 1, If $t \geq Iter_{max}$ or if the stop criteria is satisfied the algorithm finishes the iteration process and jump to step 11. Otherwise jump to step 7.

Step 11: Select the particle that has the best x_t^{Bc} objective function value (using Eqs. 4.10 or 4.16).

Step 12: Apply the thresholds values contained in x_t^{Bc} to the image I Eq. (4.2).

4.3.3 EMO Implementation with Tsallis Entropy

The proposed segmentation algorithm has been implemented considered the Tsallis pseudo-additive entropic rule as objective function (Eq. 4.25). The implementation of such algorithm can be summarized into the following steps:

Step 1:   Read the image I and store it into I_{Gr}.

Step 2:   Obtain histogram h^{Gr} of I_{Gr}.

Step 3:   Initialize the EMO parameters: $Iter_{max}$, $Iter_{local}$, δ, k and N.

Step 4:   Initialize a population $\mathbf{Sp}_t$ of N random particles with nt dimensions.

Step 5:   Compute the Tsallis entropy $S_q^i(\mathbf{Sp}_t)$ for each element of $\mathbf{Sp}_t$, Eqs. (4.26) and (4.27). Evaluate $\mathbf{Sp}_t$ in the objective function $TH_{opt}(\mathbf{Sp}_t)$ Eq. (4.25).

Step 6:   Compute the charge of each particle and the total force vector according with the method described in Chap. 3.

Step 7:   Move the entire population $\mathbf{Sp}_t$ along the total force vector.

Step 8:   Apply the local search to the moved population and select the best elements of this search based on their objective function values.

Step 9:   The t index is increased in 1, If $t \geq Iter_{max}$ or if the stop criteria is satisfied the algorithm finishes the iteration process and jump to step 11. Otherwise jump to step 7.

Step 10: Select the particle that has the best $x_t^{B^c}$ objective function value.

Step 11: Apply the thresholds values contained in $x_t^{B^c}$ to the image I_{Gr} Eq. (4.2).

4.4 Experimental Results

The aim of this section is test the segmentation algorithms based on EMO. As was previously mentioned, the most used methods are Otsu and Kapur, for that reason they are tested exhaustively in this chapter. Meanwhile, the implementation based on Tsallis entropy is tested just to verify their segmentation performance. Considering such factors, this section is divided in two main subsections: in the first one is analysed the performance of Otsu's and Kapur's methods using EMO and their segmentation capabilities are tested considering contaminated images. In the second part, the EMO algorithm considers the Tsallis entropy as objective function and it performance is tested considering a reduced number of images. In future works the Tsallis entropy will be tested with images with more complex conditions.

4.4.1 Otsu and Kapur Results

The presented algorithm based on EMO and Otsu's and Kapur's methods, has been tested under a set of 0 benchmark images. Some of these images are widely used in the image processing literature to test different methods (Lena, Cameraman, Hunter, Baboon, etc.) [20, 22]. All the images have the same size (512 × 512 pixels) and they are in JPGE format.

In order to carry out the algorithm analysis the presented MTEMO is compared to state-of-the-art thresholding methods, such Genetic Algorithms (GA) [18, 45], Particle Swarm Optimization (PSO) [3], Bacterial Foraging (BF) [22]. Since all the methods are stochastic, it is necessary to employ statistical metrics to compare the efficiency of the algorithms. Hence, all algorithms are executed 35 times per image, according to the related literature the number of the thresholds for test are $th = 2, 3, 4, 5$ [1, 3, 4]. In each experiment the stop criteria is set to 50 iterations. In order to verify the stability at the end of each test the standard deviation (STD) is obtained (Eq. 4.29). If the STD value increases the algorithms becomes more instable [1].

$$STD = \sqrt{\sum_{i=1}^{Iter_{max}} \frac{(\sigma_i - \mu)}{Ru}} \qquad (4.29)$$

On the other hand, the peak-to-signal ratio (PSNR) is used to compare the similarity of an image (image segmented) against a reference image (original image) based on the mean square error (MSE) of each pixel [4, 20, 46]. Both PSNR and MSE are defined as:

$$PSNR = 20 \log_{10} \left(\frac{255}{RMSE} \right), \quad (dB)$$

$$RMSE = \sqrt{\frac{\sum_{i=1}^{ro} \sum_{j=1}^{co} \left(\mathbf{I}_o^c(i,j) - \mathbf{I}_{th}^c(i,j) \right)}{ro \times co}}$$

(4.30)

where I_o^c is the original image, I_{th}^c is the segmented image, c depends of the image (RGB or gray scale) and ro, co are the total number of rows and columns of the image, respectively. Table 4.1 presents the parameters for the EMO algorithm. They have been obtained using the criterion proposed in [32] and kept for all test images. The EMO configuration values are selected according [32, 37] where the values are used depending on the problem dimension. In this paper the maximum number of dimension is 5. The $Iter_{max}$ does not affect the performance of EMO, however it must be evolved during a considerable but reduced number of iterations [32] after that the fitness values is not modified. In this chapter is selected the value presented in Table 4.1, for experimental purposes but in other test the stop criterion is modified to verify the performance and convergence capabilities of the proposed method. In the experiments, the stop criterion is the number of times in which the best fitness values remains with no change. Therefore, if the fitness value for the best particle remains unspoiled in 10 % of the total number of iterations ($Iter_{max}$), then the MTHEMO is stopped.

Otsu's results

This section analyzes the results of MTEMO after considering the variance among classes (Eq. 4.10) as the objective function, just as it has been proposed by Otsu [6]. The approach is applied over the complete set of benchmark images whereas the results are registered in Tables 4.2 and 4.3. Such results present the best threshold values obtained after testing the MTEMO algorithm, considering four different threshold points $th = 2, 3, 4, 5$. Tables 4.2 and 4.3 also features the PSNR, the STD and Iteration values. From the results, it is evident that the PSNR and STD values increment their magnitude as the number of threshold points increases. Notice that the Iteration values are the number of iterations that the algorithm needs to converge.

For the sake of representation, they have been selected ten images of the set to show (graphically) the segmentation results. Figures 4.1 and 4.2 present the images selected form the benchmark set and their respective histograms which possess irregular distributions (particularly Fig. 4.1j). Under such circumstances, classical methods face great difficulties to find the best threshold values.

Table 4.1 EMO parameters

$Iter_{max}$	$Iter_{local}$	δ	N
200	10	0.025	50

Table 4.2 Results after apply the MTEMO to the set of benchmark images using Otsu as objective function

Image	k	Thresholds x_t^B	PSNR	STD	Iterations
Camera man	2	70, 144	17.247	1.40 E−12	13
	3	58, 118, 155	20.226	3.07 E−01	21
	4	42, 95, 140, 170	21.533	8.40 E−03	25
	5	35, 82, 122, 149, 173	22.391	2.12 E+00	28
Lena	2	91, 149	15.480	0.00 E+00	10
	3	79, 125, 169	17.424	2.64 E−02	17
	4	73, 112, 144, 179	18.763	1.76 E−02	24
	5	71, 107, 135, 159, 186	19.442	6.64 E−01	26
Baboon	2	97, 149	15.422	6.92 E−13	15
	3	85, 125, 161	17.709	7.66 E−01	25
	4	71, 105, 136, 167	20.289	2.65 E−02	11
	5	66, 97, 123, 147, 173	21.713	4.86 E−02	22
Hunter	2	51, 116	17.875	2.31 E−12	12
	3	36, 86, 135	20.350	2.22 E−02	19
	4	27, 65, 104, 143	22.203	1.93 E−02	25
	5	23, 54,88, 112, 152	23.723	1.60 E−03	30
Airplane	2	114, 174	15.033	2.65 E−02	14
	3	92, 144, 190	18.854	9.29 E−02	28
	4	85, 130, 173, 203	20.717	1.05 E−02	26
	5	68, 106, 142, 179, 204	23.160	2.38 E−02	31
Peppers	2	72, 138	16.299	1.38 E−12	16
	3	65, 122, 169	18.359	4.61 E−13	20
	4	50, 88, 128, 171	20.737	4.61 E−13	25
	5	48, 85, 118, 150, 179	22.310	2.33 E−02	34
Living room	2	87, 145	15.999	1.15 E−12	18
	3	76, 123, 163	18.197	6.92 E−12	24
	4	56, 97, 132, 168	20.673	1.78 E−01	29
	5	49, 88, 120, 147, 179	22.192	1.02 E−01	28
Blonde	2	106, 155	14.609	3.70 E−03	15
	3	53, 112, 158	19.157	9.23 E−13	20
	4	50, 103, 139, 168	20.964	2.53 E−02	29
	5	48, 95, 125, 151, 174	22.335	4.50 E−02	32
Bridge	2	91, 56	13.943	4.61 E−13	11
	3	72, 120, 177	17.019	1.11 E+00	16
	4	63, 103, 145, 193	18.872	3.20 E−01	17
	5	59, 95, 127, 161, 291	20.143	7.32 E−01	27
Butterfly	2	99, 151	13.934	9.68 E−02	10
	3	82, 119, 160	16.932	1.15 E−12	15
	4	81, 114, 145, 176	17.323	3.38 E+00	33
	5	61, 83, 106, 130, 163	21.683	2.86 E+00	25
Lake	2	86, 155	14.647	2.53 E−02	18
	3	79, 141, 195	15.823	3.99 E−02	24
	4	67, 111, 159, 199	17.642	3.91 E−02	32
	5	57, 88, 127, 166, 200	19.416	4.89 E−02	40

Table 4.3 Results after apply the MTEMO to the set of benchmark images using Otsu as objective function

Image	k	Thresholds x_i^B	PSNR	STD	Iterations
Arch monument	2	70, 143	15.685	2.20 E−03	10
	3	49, 96, 156	18.257	2.50 E−03	27
	4	42, 80, 126, 174	20.190	1.78 E−02	24
	5	36, 67, 101, 141, 183	21.738	7.15 E−02	20
Firemen	2	61, 145	15.511	9.22 E−13	10
	3	45, 96, 161	17.919	2.20 E−02	21
	4	43, 88, 139, 191	19.832	1.41 E−02	28
	5	38, 75, 108, 152, 198	21.266	4.53 E−02	15
Maize	2	91, 167	13.853	1.65 E−02	14
	3	76, 128, 187	15.537	2.16 E−02	18
	4	66, 106, 152, 201	16.972	1.58 E−02	15
	5	58, 89, 126, 166, 209	18.476	5.75 E−02	53
Native fisherman	2	107, 196	12.630	9.22 E−13	15
	3	88, 135, 206	15.015	6.90 E−03	12
	4	67, 105, 144, 209	17.571	2.49 E−02	26
	5	62, 96, 126, 157, 214	18.835	2.78 E−12	20
Pyramid	2	114, 167	12.120	4.61 E−13	16
	3	96, 129, 175	15.765	5.55 E−02	16
	4	90, 119, 146, 186	17.437	1.98 E−02	40
	5	86, 111, 133, 158, 195	18.582	4.11 E−02	26
Sea star	2	85, 157	14.815	4.61 E−13	15
	3	68, 119, 177	17.357	5.90 E−03	11
	4	60, 101, 138, 187	19.125	5.11 E−02	44
	5	52, 86, 117, 150, 194	20.729	5.75 E−02	12
Smiling girl	2	66, 139	16.783	4.61 E−13	11
	3	61, 127, 162	18.827	1.30 E−03	20
	4	55, 111, 143, 171	21.137	5.80 E−02	33
	5	47, 97, 128, 154, 178	23.221	4.28 E−02	27
Surfer	2	93, 163	12.490	15.7 E−02	22
	3	71, 110, 176	15.983	6.92 E−13	21
	4	47, 81, 118, 181	20.677	2.40 E−03	45
	5	46, 77, 106, 143, 197	21.864	5.84 E−02	27
Train	2	91, 75	14.341	0.00 E+00	12
	3	61, 118, 179	18.141	1.38 E−12	14
	4	55, 106, 142, 187	20.050	4.61 E−13	26
	5	54, 104, 138, 170, 211	21.112	2.03 E+00	25

Tables 4.4 and 4.5 show the images obtained after processing 10 original images selected from the entire benchmark set, applying the proposed algorithm. The results present the segmented images considering four different threshold points

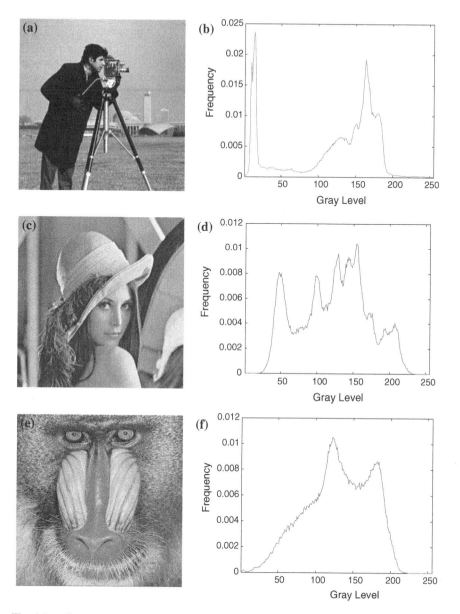

Fig. 4.1 **a** Camera man, **c** Lena, **e** Baboon, **g** Hunter and **i** Butterfly, the selected benchmark images. **b**, **d**, **f**, **h**, **j** histograms of the images

$th = 2, 3, 4, 5$. In Tables 4.4 and 4.5, it is also shown the evolution of the objective function during one execution. From the results, it is possible to appreciate that the MTEMO converges (stabilizes) around the first 50 iterations. However the algorithm continues running in order to show the convergence capabilities. The

Fig. 4.1 (continued)

segmented images provide evidence that the outcome is better with $th = 4$ and $th = 5$; however, if the segmentation task does not requires to be extremely accurate then it is possible to select $th = 3$.

Kapur's results

This section analyses the performance of the proposed approach based on EMO considering as objective function (Eq. 4.16) the entropy function proposed by Kapur [7], is analyzed. In Tables 4.6 and 4.7, are presented the experimental results after the application of MTEMO over the entire set of benchmark images. The values listed are *PSNR*, *STD*, Iterations and the best threshold values of the last population (x_t^B).

Tables 4.8 and 4.9 show the images obtained after processing 10 original images selected from the entire benchmark set, applying the presented algorithm. The results present the segmented images considering four different threshold points $th = 2, 3, 4, 5$.

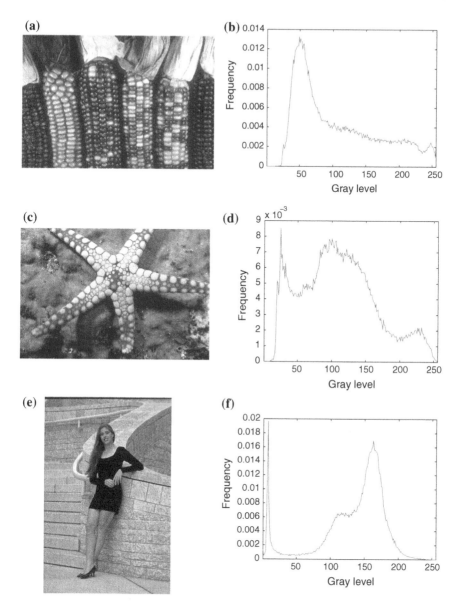

Fig. 4.2 **a** Maize, **c** Sea star, **e** Smiling girl, **g** Surfer and **i** Train, the selected benchmark images. **b**, **d**, **f**, **h**, **j** histograms of the images

EMO using Otsu's and Kapur's methods over contaminated images

Another important test consist in add two different kind of noise to a selected test image. The objective is to verify if the proposed algorithm is able to segment the contaminated images. Gaussian noise is used in this test; its parameters are $\mu = 0$

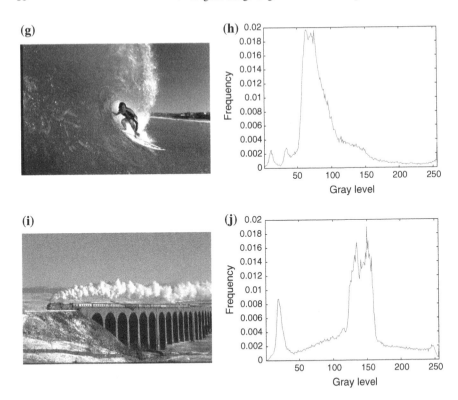

Fig. 4.2 (continued)

(mean) and $\sigma = 0.1$ (variance). On the other hand, a 2 % of Salt and Pepper (impulsive) noise is used to contaminate the image. Both Gaussian and Salt and Pepper noise are generated using the corresponding Matlab functions.

Figure 4.3 presents the original Train image taken from the entire benchmark set. Moreover the noised images are presented and their respective histograms. Such histograms show that they are modified when the original image is noised. Obviously the pixels are contaminated and their distribution is affected. However in Table 4.10 are presented the results after apply the proposed method with Otsu's function over the noised images.

Table 4.11 presents the experimental results after apply the Kapur's proposed algorithm over the noised Train image, for four different th values ($th = 2, 3, 4, 5$).

Comparisons EMO using Otsu's and Kapur's functions

In order to analyse the results of the proposed approach based on EMO and Otsu's and Kapur's objective functions, three different comparisons are executed. The first one involves the comparison between the two versions of MTEMO, one with the Otsu function and the other with the Kapur criterion. The second one analyses the comparison among the MTEMO with other state-of-the-art

Table 4.4 Results after apply the MTEMO using Otsu's over the selected benchmark images

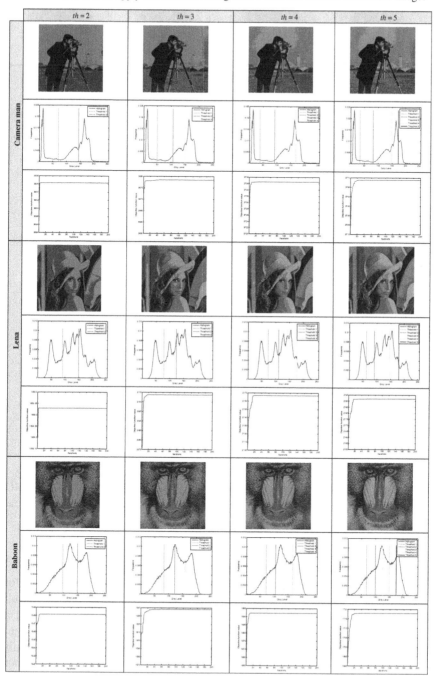

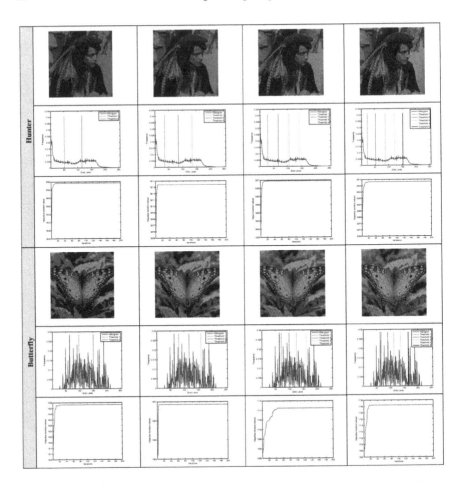

approaches. Finally the third one compares the number of iterations of MTEMO and the selected methods, in order to verify its performance and computational effort.

Comparisons between Otsu and Kapur

In order to statistically compare the results from Tables 4.2, 4.3, 4.6 and 4.7, a non-parametric significance proof known as the Wilcoxon's rank test [47, 48] for 35 independent samples has been conducted. Such proof allows assessing result differences among two related methods. The analysis is performed considering a 5 % significance level over the peak-to-signal ratio (PSNR) data corresponding to the five threshold points. Table 4.12 reports the p-values produced by Wilcoxon's test for a pair-wise comparison of the PSNR values between the Otsu and Kapur objective functions. As a null hypothesis, it is assumed that there is no difference between the values of the two objective functions. The alternative hypothesis

Table 4.5 Results after apply the MTEMO using Otsu's over the selected benchmark images

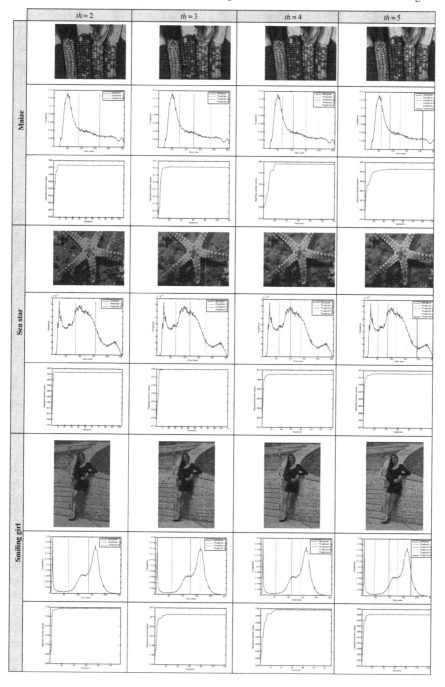

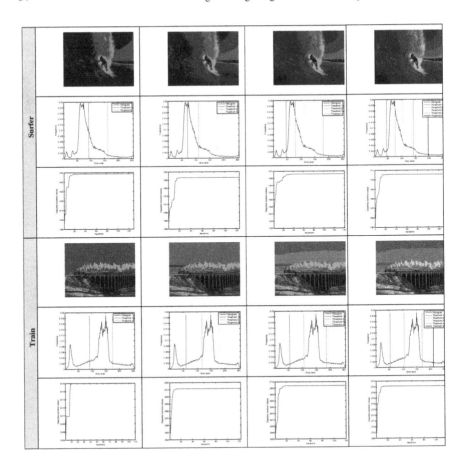

considers an existent difference between the values of both approaches. All *p*-values reported in Table 4.12 are less than 0.05 (5 % significance level) which is a strong evidence against the null hypothesis, indicating that the Otsu PSNR mean values for the performance are statistically better and it has not occurred by chance.

Comparison among EMO based algorithm and other MT approaches

In order to demonstrate that the MTEMO is an interesting alternative for MT, the proposed algorithm is compared with other similar implementations. The other methods used in the comparison are: Genetic Algorithms (GA), Particle Swarm Optimization (PSO) and Bacterial foraging (BF).

All the algorithms run 35 times over each selected image. The images used for this test are the same of the selected in previous experiments (Camera man, Lena, Baboon, Hunter, Butterfly, Maize, Sea star, Smiling girl, Surfer and Train). For each image is computed the *PSNR*, *STD* and the mean of the objective function

Table 4.6 Results after apply the MTEMO to the set of benchmark images using Kapur as objective function

Image	k	Thresholds x_i^B	PSNR	STD	Iterations
Camera man	2	128, 196	13.626	3.60 E−15	18
	3	97, 146, 196	18.803	4.91 E−02	25
	4	44, 96, 146, 196	20.586	1.08 E−14	29
	5	24, 60, 98, 146, 196	20.661	6.35 E−02	27
Lena	2	95, 163	14.672	0.00 E+00	18
	3	81, 126, 176	17.247	7.50 E−04	25
	4	76, 118, 158, 190	18.251	1.34 E−02	33
	5	61, 92, 126, 161, 192	20.019	2.67 E−02	27
Baboon	2	79, 143	16.016	1.08 E−14	19
	3	79, 143, 231	16.016	3.60 E−15	38
	4	44, 98, 152, 231	18.485	2.10 E−03	22
	5	33, 74, 114, 159, 231	20.507	1.08 E−14	25
Hunter	2	92, 179	15.206	1.44 E−14	17
	3	59, 127, 179	18.500	4.82 E−04	23
	4	44, 89, 133, 179	21.728	3.93 E−04	20
	5	46, 90, 133, 179, 222	21.073	4.20 E−02	28
Airplane	2	70, 171	15.758	3.30 E−03	18
	3	68, 126, 182	18.810	1.08 E−14	23
	4	68, 126, 182, 232	18.810	2.37 E−01	30
	5	64, 105, 145, 185, 232	20.486	1.87 E−01	32
Peppers	2	66, 143	16.265	7.21 E−15	15
	3	62, 112, 162	18.367	2.80 E−03	21
	4	62, 112, 162, 227	18.376	1.28 E−01	29
	5	48, 86, 127, 171, 227	20.643	1.37 E−01	32
Living room	2	89, 170	14.631	2.43 E−04	19
	3	47, 103, 175	17.146	1.08 E−10	25
	4	47, 102, 153, 197	19.068	8.90 E−03	23
	5	42, 81, 115, 158, 197	21.155	1.00 E−02	28
Blonde	2	125, 203	12.244	1.83 E−01	16
	3	65, 134, 203	16.878	1.40 E−01	24
	4	65, 113, 155, 203	20.107	1.95 E−01	26
	5	65, 100, 134, 166, 203	22.138	1.01 E−01	29
Bridge	2	94, 171	13.529	1.05 E−02	18
	3	65, 131, 195	16.806	1.08 E−10	19
	4	53, 102, 151, 199	18.902	1.44 E−14	26
	5	36, 73, 114, 159, 203	20.733	1.75 E−03	24
Butterfly	2	120, 213	11.065	1.35 E−01	22
	3	96, 144, 213	14.176	3.56 E−01	29
	4	27, 96, 144, 213	16.725	3.45 E−01	36
	5	27, 85, 120, 152, 213	19.026	2.32 E−01	30

(continued)

Table 4.6 (continued)

Image	k	Thresholds x_i^B	PSNR	STD	Iterations
Lake	2	91, 163	14.713	1.44 E−14	19
	3	73, 120, 170	16.441	9.55 E−05	23
	4	69, 112, 156, 195	17.455	1.73 E−02	25
	5	62, 96, 131, 166, 198	18.774	5.45 E−02	36

Table 4.7 Results after apply the MTEMO to the set of benchmark images using Otsu as objective function

Image	k	Thresholds x_i^B	PSNR	STD	Iterations
Arch	2	80, 155	15.520	1.44 E−14	8
	3	64, 118, 174	17.488	1.80 E−14	29
	4	61, 114, 165, 215	17.950	5.03 E−02	69
	5	48, 89, 130, 172, 217	20.148	5.93 E−02	17
Firemen	2	102, 175	14.021	6.29 E−04	22
	3	72, 127, 184	17.146	2.90 E−02	10
	4	70, 123, 172, 220	17.782	4.78 E−05	33
	5	53, 92, 131, 176, 221	20.572	1.44 E−14	40
Maize	2	98, 176	13.633	2.26 E−04	25
	3	81, 140, 198	15.229	5.92 E−05	24
	4	74, 120, 165, 211	16.280	2.97 E−04	38
	5	68, 105, 143, 180, 218	17.211	3.27 E−04	40
Native fisherman	2	68, 154	11.669	7.20 E−15	23
	3	52, 122, 185	14.293	1.15 E−02	21
	4	48, 100, 150, 197	16.254	1.44 E−14	18
	5	38, 73, 113, 154, 198	17.102	1.24 E−02	33
Pyramid	2	36, 165	10.081	0.00 E+00	24
	3	36, 110, 173	15.843	0.00 E+00	15
	4	36, 98, 158, 199	17.256	3.05 E−02	24
	5	36, 88, 124, 161, 201	20.724	5.71 E−02	20
Sea star	2	90, 169	14.398	7.20 E−15	23
	3	75, 130, 184	16.987	1.08 E−14	20
	4	67, 115, 163, 206	18.304	5.02 E−04	40
	5	56, 94, 133, 172, 211	20.165	7.51 E−04	45
Smiling girl	2	106, 202	13.420	0.00 E+00	17
	3	94, 143, 202	18.254	6.06 E−05	22
	4	36, 84, 139, 202	18.860	1.96 E−02	20
	5	36, 84, 134, 178, 211	19.840	5.42 E−02	22
Surfer	2	105, 172	11.744	1.02 E−02	22
	3	51, 106, 172	18.584	7.49 E−02	32
	4	51, 102, 155, 203	19.478	3.06 E−15	28
	5	51, 97, 136, 172, 213	20.468	6.50 E−03	24
Train	2	105, 169	14.947	0.00 E+00	18
	3	70, 120, 171	18.212	8.20 E−03	18
	4	70, 120, 162, 208	19.394	1.44 E−14	26
	5	39, 79, 121, 162, 208	20.619	4.56 E−02	24

Table 4.8 Results after applying the MTEMO using Kapur's over the selected benchmark images

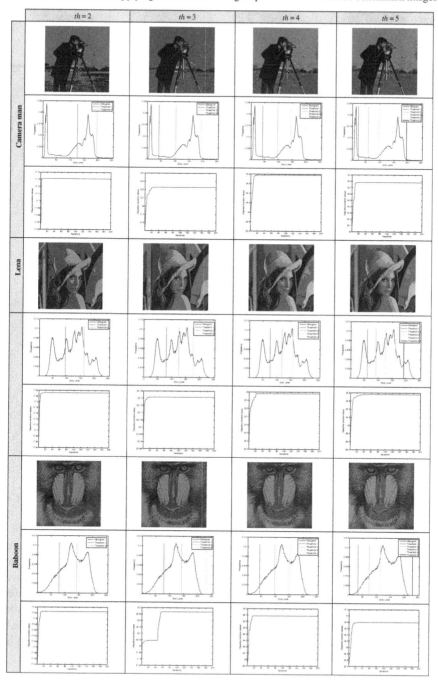

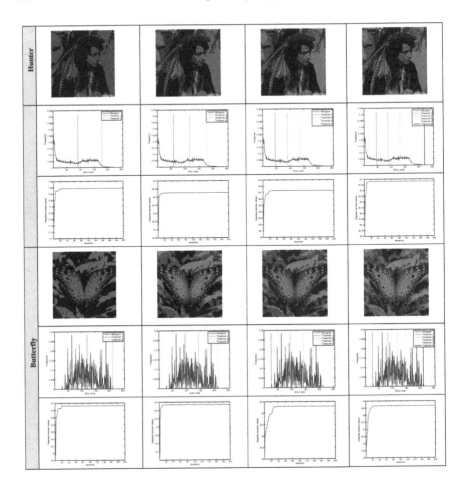

values, moreover the entire test is performed using both Otsu's and Kapur's objective functions.

Table 4.13 presents the computed values for the reduced benchmark test (ten images), the values in bold represent the best values founded at the end of the entire test. It is possible to see how the MTEMO algorithm has better performance than the others. Such values are computed using the Otsu's method as a objective function. On the other hand, the same experiment has been performed using the Kapur's method. Using the same criteria described for the Otsu's method the algorithm runs over 35 times in each image. The results of this experiment are presented in Table 4.14. The results show that the proposed MTEMO algorithm is better in comparison with the GA, PSO and BF.

Table 4.9 Results after apply the MTEMO using Kapur's over the selected benchmark images

Performance and computational effort compared with other approaches

In this subsection is compared the performance and computational effort of the proposed method and the GA, PSO and BF approaches. Table 4.15 presents the number of iterations required for each algorithm to achieve a stable objective function value. There are used both Otsu's and Kapur's functions to find the best threshold values in each one of the ten images selected from the entire set of test images.

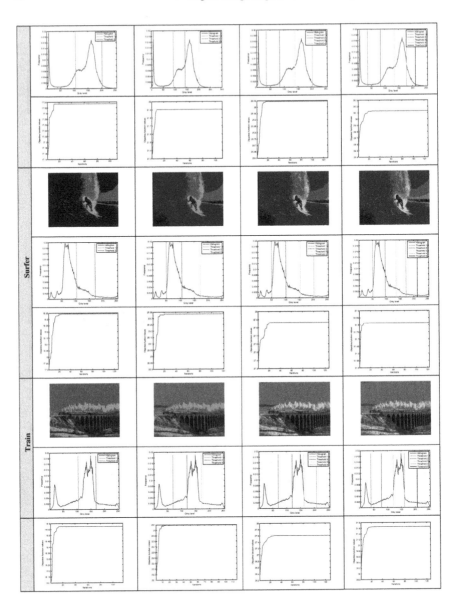

The iterations values of Table 4.15 provide evidence that the MTEMO requires less iteration to find a stable value. In [32] is provided a proof that EMO requires a low number of iterations depending on the dimension of the problem. Based on this the obtained confirm that the computational effort of MTEMO is better than GA, PSO and BF for multilevel thresholding. On other hand, the iteration values are tested using the non-parametric Wilcoxon ranking test. The test is divided in three

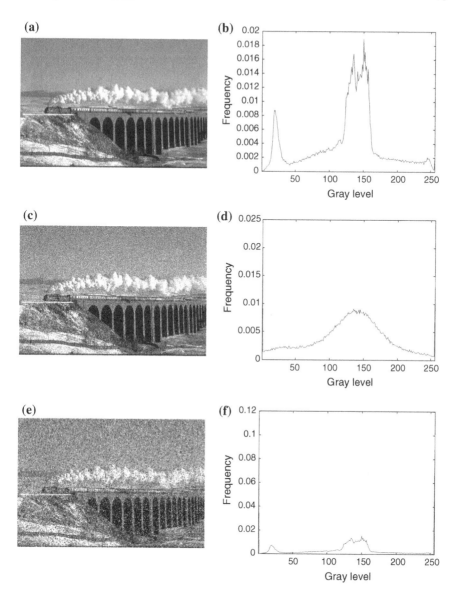

Fig. 4.3 a Original Train image, **c** Gaussian contaminated Train image, **e** Salt and pepper contaminated Train image, **b, d, f** histograms of the images

groups MTEMO versus GA, MTEMO versus PSO and MTEMO versus BF, the *p*-values of Wilcoxon are presented in Table 4.16.

The results reported on Table 4.16 are less than 0.05 (5 % significance level) which is strong evidence against the null hypothesis, this indicates that the iteration values of MTEMO are statistically better and it they have not occurred by a chance.

Table 4.10 Results after apply the MTEMO using Otsu's over the noised Train image

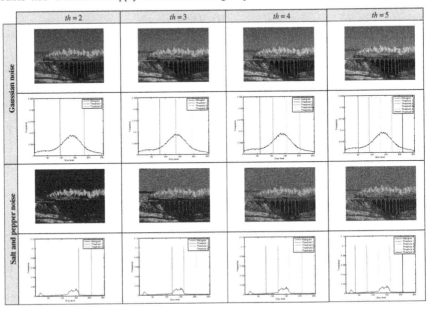

Table 4.11 Results after apply the MTEMO using Kapur's over the noised Train image

Table 4.12 p-values produced by Wilcoxon's test comparing Otsu versus Kapur over the averaged PSNR from Tables 4.2, 4.3, 4.6 and 4.7

Image	p-Value Otsu versus Kapur
Camera man	2.8061e−005
Lena	1.2111e−004
Baboon	2.6722e−004
Hunter	2.1341e−004
Airplane	8.3241e−005
Peppers	7.9341e−005
Living room	1.4522e−004
Blonde	9.7101e−005
Bridge	1.3765e−004
Butterfly	6.2955e−005
Lake	4.7303e−005
Arch	4.9426e−005
Firemen	4.7239e−005
Maize	1.6871e−004
Native fisherman	3.5188e−004
Pyramid	9.3876e−005
Sea star	1.4764e−005
Smiling girl	7.1464e−004
Surfer	9.5993e−005
Train	3.5649e−004

4.4.2 Tsallis Entropy Results

In this section, the results of the TSEMO algorithm are analyzed, considering as objective function (Eq. 4.25) the Tsallis entropy [15]. The approach is applied over the complete set of benchmark images whereas the results are registered in Table 4.17. Such results present the best threshold values obtained after testing the TSEMO algorithm, considering four different threshold points $th = 2, 3, 4, 5$. In Table 4.17, it is also shown the *PSNR*, *STD*, *SSIM* and *FSIM* values.

There have been selected five images of the set to show (graphically) the segmentation results. The selected images are presented in Fig. 4.1. Table 4.18 shows the images obtained after processing 5 original images selected from the entire benchmark set, applying the proposed algorithm. The results present the segmented images considering four different threshold levels $th = 2, 3, 4, 5$. In Table 4.18, it is also shown the evolution of the objective function during one execution. From the results, it is possible to appreciate that the TSEMO converges (stabilizes) around the first 100 iterations. The segmented images provide evidence that the outcome is better with $th = 4$ and $th = 5$; however, if the segmentation task does not requires to be extremely accurate then it is possible to select $th = 3$.

Table 4.13 Comparisons between MTEMO, GA, PSO and BF, applied over the selected test images using Otsu's method

Image	k	MTEMO			GA			PSO			BF		
		PSNR	STD	Mean	PSNR	STD	Mean	PSNR	STD	Mean	PSNR	STD	Mean
Camera man	2	17.247	1.40 E-12	3606.3	17.048	0.0232	3604.5	17.033	0.0341	3598.3	17.058	0.0345	3590.9
	3	20.226	3.07 E-01	3679.5	17.573	0.1455	3678.3	19.219	0.2345	3662.7	20.035	0.2459	3657.5
	4	21.533	8.40 E-03	3782.4	20.523	0.2232	3781.5	21.254	0.3142	3777.4	21.209	0.4560	3761.4
	5	22.391	2.12 E+00	3767.6	21.369	0.4589	3766.4	22.095	0.5089	3741.6	22.237	0.5089	3789.8
Lena	2	15.480	0.00 E+00	1939.3	15.040	0.0049	1960.9	15.077	0.0033	1961.4	15.031	2.99 E-04	1961.5
	3	17.424	2.64 E-02	2103.8	17.304	0.1100	2126.4	17.276	0.0390	2127.7	17.401	0.0061	2128.0
	4	18.763	1.76 E-02	2166.8	17.920	0.2594	2173.7	18.305	0.1810	2180.6	18.507	0.0081	2189.0
	5	19.442	6.64 E-01	2192.4	18.402	0.3048	2196.2	18.770	0.2181	2212.5	19.001	0.0502	2215.6
Baboon	2	15.422	6.92 E-13	1548.1	15.304	0.0031	1547.6	15.088	0.0077	1547.9	15.353	8.88 E-04	1548.0
	3	17.709	7.66 E-01	1638.3	17.505	0.1750	1633.5	17.603	0.0816	1635.3	17.074	0.0287	1637.0
	4	20.289	2.65 E-02	1692.1	18.708	0.2707	1677.7	19.233	0.0853	1684.3	19.654	0.0336	1690.7
	5	21.713	4.86 E-02	1717.8	20.203	0.3048	1712.9	20.526	0.1899	1712.9	21.160	0.1065	1716.7
Hunter	2	17.875	2.31 E-12	3064.2	17.088	0.0470	3064.1	17.932	0.2534	3064.1	17.508	0.0322	3064.1
	3	20.350	2.22 E-02	3213.4	20.045	0.1930	3212.9	19.940	0.9727	3212.4	20.350	0.9627	3213.4
	4	22.203	1.93 E-02	3269.5	20.836	0.6478	3268.4	21.128	2.2936	3266.3	21.089	2.2936	3266.3
	5	23.723	1.60 E-03	3308.1	21.284	1.6202	3305.6	22.026	4.1811	3276.3	22.804	3.6102	3291.1
Butterfly	2	13.934	9.68 E-02	1553.0	13.007	0.0426	1553.0	13.092	0.0846	1553.0	13.890	0.0643	1553.0
	3	16.932	1.15 E-12	1669.3	15.811	0.3586	1669.0	17.261	2.6268	1665.7	17.285	1.2113	1667.2
	4	17.323	3.38 E+00	1709.1	17.104	0.6253	1709.9	17.005	3.7976	1702.9	17.128	2.2120	1707.0
	5	21.683	2.86 E+00	1735.0	18.593	0.5968	1734.4	18.099	6.0747	1730.7	18.9061	3.5217	1733.0
Maize	2	13.853	1.65 E-02	3562.7	13.014	0.0257	3500.5	13.693	6.3521	3560.7	13.712	0.0781	3459.9
	3	15.537	2.16 E-02	3720.2	15.112	0.1538	3699.7	15.008	21.504	3712.2	15.200	0.2789	3701.0
	4	16.972	1.58 E-02	3799.1	16.203	0.3287	3701.5	16.157	17.521	3790.9	16.781	0.3681	3750.8
	5	18.476	5.75 E-02	3843.1	17.953	0.8569	3799.9	17.740	14.787	3836.2	18.102	0.7163	3810.0

(continued)

Table 4.13 (continued)

Image	k	MTEMO			GA			PSO			BF		
		PSNR	STD	Mean	PSNR	STD	Mean	PSNR	STD	Mean	PSNR	STD	Mean
Sea star	2	14.815	4.61 E−13	2546.9	14.744	0.0879	2534.8	14.802	3.0898	2345.6	14.798	0.0091	2352.8
	3	17.357	5.90 E−03	2779.9	17.034	0.1236	2699.8	17.339	11.582	2676.3	17.330	0.0398	2720.8
	4	19.125	5.11 E−02	2865.7	18.482	0.1897	2820.1	18.112	19.070	2657.5	18.818	0.2651	2821.9
	5	20.729	5.75 E−02	2912.8	19.383	0.3647	2903.0	19.019	19.083	2890.4	20.760	1.8793	2895.6
Smiling girl	2	16.783	4.61 E−13	2107.8	16.248	0.0129	2103.9	16.701	0.6896	2067.1	16.548	0.0359	2105.0
	3	18.827	1.30 E−03	2211.5	18.157	0.2987	2190.0	18.800	4.4323	2200.2	18.756	0.1569	2110.3
	4	21.137	5.80 E−02	2264.3	18.816	0.7964	2250.9	20.323	11.076	2250.3	21.091	0.3952	2259.8
	5	23.221	4.28 E−02	2295.5	19.219	1.9871	2279.7	22.628	9.7178	2285.1	22.980	2.7816	2281.3
Surfer	2	12.490	15.7 E−02	1448.6	12.001	0.0373	1342.5	12.579	1.7211	1448.0	12.109	0.0449	1395.6
	3	15.983	6.92 E−13	1586.5	14.509	0.1782	1456.7	14.789	1.7653	1586.1	15.900	0.3890	1487.6
	4	20.677	2.40 E−03	1665.9	19.987	0.3513	1569.8	19.965	14.787	1659.4	19.992	0.5790	1598.7
	5	21.864	5.84 E−02	1705.9	20.892	0.4789	16.001	21.575	13.274	1699.1	20.980	1.1239	1690.0
Train	2	14.341	0.00 E+00	2418.0	13.986	0.0138	2407.5	13.933	4.1810	2416.6	14.292	0.0069	2416.9
	3	18.141	1.38 E−12	2611.5	17.471	0.2715	2604.6	17.947	18.797	2606.5	17.992	0.1450	2610.8
	4	20.050	4.61 E−13	2697.0	18.082	0.3819	2661.4	19.131	12.443	2691.9	19.796	0.7283	2684.2
	5	21.112	2.03 E+00	2740.3	20.303	0.4418	2726.6	20.997	12.719	2732.7	20.778	0.7404	2727.1

Table 4.14 Comparisons between MTEMO, GA, PSO and BF, applied over the selected test images using Kapur's method

Image	k	MTEMO			GA			PSO			BF		
		PSNR	STD	Mean	PSNR	STD	Mean	PSNR	STD	Mean	PSNR	STD	Mean
Camera man	2	13.626	3.60 E−15	17.584	11.941	0.1270	15.341	12.259	0.1001	16.071	12.264	0.0041	16.768
	3	18.803	4.91 E−02	21.976	14.827	0.2136	20.600	15.211	0.1107	21.125	15.250	0.0075	21.498
	4	20.586	1.08 E−14	26.586	17.166	0.2857	24.267	18.000	0.2005	25.050	18.406	0.0081	25.093
	5	20.661	6.35 E−02	30.506	19.795	0.3528	28.326	20.963	0.2734	28.365	21.211	0.0741	30.026
Lena	2	14.672	0.00 E+00	17.831	12.334	0.0049	16.122	12.345	0.0033	16.916	12.345	2.99 E-4	16.605
	3	17.247	7.50 E−04	22.120	14.995	0.1100	20.920	15.133	0.0390	20.468	15.133	0.0061	20.812
	4	18.251	1.34 E−02	25.999	17.089	0.2594	23.569	17.838	0.1810	24.449	17.089	0.0081	26.214
	5	20.019	2.67 E−02	29.787	19.549	0.3043	27.213	20.442	0.2181	27.526	19.549	0.0502	28.046
Baboon	2	16.016	1.08 E−14	17.625	12.184	0.0567	16.425	12.213	0.0077	16.811	12.216	8.88 E-4	16.889
	3	16.016	3.60 E−15	22.269	14.745	0.1580	21.069	15.008	0.0816	21.088	15.211	0.0287	21.630
	4	18.485	2.10 E−03	26.688	16.935	0.1765	25.489	17.574	0.0853	24.375	17.999	0.0336	25.446
	5	20.507	1.08 E−14	30.800	19.662	0.2775	29.601	20.224	0.1899	30.994	20.720	0.1065	30.887
Hunter	2	15.206	1.44 E−14	17.856	12.349	0.0148	16.150	12.370	0.0068	15.580	12.373	0.0033	16.795
	3	18.500	4.82 E−04	22.525	14.838	0.1741	21.026	15.128	0.0936	20.639	15.553	0.1155	21.860
	4	21.729	3.93 E−04	26.728	17.218	0.2192	25.509	18.040	0.1560	27.085	18.381	0.0055	26.230
	5	21.074	4.20 E−02	30.642	19.563	0.3466	29.042	20.533	0.2720	29.013	21.256	0.0028	28.856
Butterfly	2	11.0653	1.35 E−01	16.681	10.470	0.0872	15.481	10.474	0.0025	14.098	10.474	0.0014	15.784
	3	14.1766	3.56 E−01	21.242	11.628	0.2021	20.042	12.313	0.1880	19.340	12.754	0.0118	21.308
	4	16.7257	3.45 E−01	25.179	13.314	0.2596	23.980	14.231	0.2473	25.190	14.877	0.0166	25.963
	5	19.0267	2.32 E−01	28.611	15.756	0.3977	27.411	16.337	0.2821	27.004	16.828	0.0877	27.980
Maize	2	13.633	2.26 E−04	18.631	13.506	0.0725	18.521	13.466	0.0012	18.631	13.601	0.0022	18.625
	3	15.229	5.92 E−05	23.565	15.150	0.1582	23.153	15.018	0.0530	23.259	15.032	0.0068	23.128
	4	16.280	2.97 E−04	27.529	15.909	0.2697	26.798	15.834	0.1424	27.470	16.120	0.0128	27.198
	5	17.211	3.27 E−04	31.535	16.921	0.8971	30.852	16.319	0.4980	31.255	16.985	0.0978	30.987

(continued)

Table 4.14 (continued)

Image	k	MTEMO			GA			PSO			BF		
		PSNR	STD	Mean	PSNR	STD	Mean	PSNR	STD	Mean	PSNR	STD	Mean
Sea star	2	14.398	7.20 E−15	18.754	14.282	0.0816	18.753	14.346	0.0002	18.593	14.280	0.0016	18.753
	3	16.987	1.08 E−14	23.323	8.2638	0.1987	23.260	16.949	0.1723	23.289	16.319	0.1813	23.292
	4	18.304	5.02 E−04	27.582	15.035	0.2691	26.533	18.389	0.2481	27.407	18.240	0.2092	26.938
	5	20.165	7.51 E−04	31.562	19.005	0.9740	30.798	19.849	0.6159	31.288	19.052	0.3553	30.857
Smiling girl	2	13.420	0.00 E+00	17.334	13.092	0.0178	17.295	13.352	0.0368	17.321	13.370	0.0038	17.272
	3	18.254	6.06 E−05	21.904	17.764	0.2179	21.580	18.201	0.0556	21.887	18.207	0.0178	21.847
	4	18.860	1.96 E−02	26.040	17.923	0.3024	25.432	18.063	0.2817	25.815	18.340	0.2119	25.183
	5	19.840	5.42 E−02	30.089	19.026	0.7128	27.940	19.200	0.5887	29.700	19.786	0.3813	28.300
Surfer	2	11.744	1.02 E−02	18.339	11.521	0.0219	18.237	11.698	0.1144	18.194	11.425	0.0489	18.269
	3	18.584	7.49 E−02	23.231	17.181	0.1715	22.865	18.413	0.2332	22.214	18.509	0.1369	23.089
	4	19.478	3.06 E−15	27.863	18.868	0.2093	26.447	19.125	0.4214	26.676	19.388	0.8240	26.859
	5	20.468	6.50 E−03	31.823	19.521	0.3182	30.363	19.491	0.4789	30.587	19.935	0.9684	30.968
Train	2	14.947	0.00 E+00	18.574	14.857	0.0222	18.573	14.933	0.0004	18.574	14.795	0.0080	18.487
	3	18.212	8.20 E−03	23.107	17.803	0.2084	22.663	18.185	0.1013	23.084	18.081	0.0772	22.009
	4	19.394	1.44 E−14	27.608	18.932	0.3065	26.510	18.667	0.4335	27.335	19.327	0.2617	26.564
	5	20.619	4.56 E−02	31.647	19.781	1.1560	30.196	20.525	0.4122	31.484	20.361	0.7846	30.688

Table 4.15 Iterations comparison between MTEMO, GA, PSO and BF, applied over the selected test images using Otsu's and Kapur's methods

Image	k	Otsu				Kapur			
		MTEMO Iterations	GA Iterations	PSO Iterations	BF Iterations	MTEMO Iterations	GA Iterations	PSO Iterations	BF Iterations
Camera man	2	13	184	132	90	18	237	93	131
	3	21	300	287	138	25	195	133	206
	4	25	535	431	129	29	315	243	254
	5	28	583	755	396	27	441	366	305
Lena	2	10	142	116	73	18	193	224	180
	3	17	314	230	152	25	280	338	265
	4	24	415	397	147	33	277	351	240
	5	26	620	386	335	27	476	422	308
Baboon	2	15	186	167	116	19	286	378	140
	3	25	348	267	180	38	368	386	275
	4	11	443	369	179	22	410	690	483
	5	22	632	518	288	25	789	755	518
Hunter	2	12	254	171	180	17	238	185	176
	3	19	278	191	74	23	264	353	187
	4	25	494	385	253	20	446	482	328
	5	30	803	406	356	28	659	884	364
Butterfly	2	10	240	173	112	22	290	300	217
	3	15	331	240	144	29	339	374	276
	4	33	341	515	297	36	462	424	304
	5	25	705	581	134	30	755	500	345
Maize	2	10	152	288	156	25	115	334	168
	3	27	188	473	178	24	145	491	198
	4	24	201	642	185	38	197	588	201
	5	20	225	921	235	40	208	811	195

(continued)

Table 4.15 (continued)

Image	k	Otsu				Kapur			
		MTEMO Iterations	GA Iterations	PSO Iterations	BF Iterations	MTEMO Iterations	GA Iterations	PSO Iterations	BF Iterations
Sea star	2	**15**	235	333	221	23	270	334	191
	3	**11**	401	440	356	20	332	540	178
	4	**44**	543	753	362	40	356	589	273
	5	**12**	606	703	470	45	496	828	315
Smiling girl	2	**11**	524	300	143	17	250	446	197
	3	**20**	472	549	269	22	340	681	341
	4	**33**	388	616	456	20	445	852	689
	5	**27**	645	723	573	22	780	992	754
Surfer	2	**22**	502	324	149	22	193	526	378
	3	**21**	431	535	193	32	235	622	493
	4	**45**	322	511	217	28	399	819	697
	5	**27**	494	950	298	24	590	793	795
Train	2	**12**	511	342	189	18	434	361	257
	3	**14**	462	431	225	18	489	474	349
	4	**26**	516	688	348	26	671	719	493
	5	**25**	599	794	458	24	719	951	544

The values in bold signifies the best values

Table 4.16 p-values produced by Wilcoxon's test comparing Otsu versus Kapur over the iteration values

Image	k	p-Value MTEMO versus GA	p-Value MTEMO versus PSO	p-Value MTEMO versus BF
Camera man	2	2.8263 E−14	4.1495 E−12	1.6185 E−14
	3	2.5482 E−15	7.1815 E−11	3.1253 E−15
	4	2.0829 E−16	1.6967 E−14	1.8069 E−13
	5	9.2180 E−16	8.3666 E−16	2.4299 E−14
Lena	2	1.9023 E−16	6.1475 E−11	2.4129 E−09
	3	5.7370 E−15	8.6537 E−14	7.9517 E−05
	4	7.9129 E−14	6.9820 E−15	1.7320 E−12
	5	3.5309 E−12	4.9352 E−13	1.9006 E−11
Baboon	2	3.4520 E−09	1.9000 E−12	2.0524 E−14
	3	9.1500 E−07	2.3250 E−06	3.6593 E−03
	4	6.8490 E−05	1.4202 E−14	9.5561 E−11
	5	3.6003 E−08	1.1213 E−14	9.9423 E−14
Hunter	2	6.1892 E−13	3.9321 E−16	8.1806 E−06
	3	4.4766 E−13	6.7790 E−15	5.4107 E−09
	4	7.4115 E−14	7.0460 E−13	5.2770 E−14
	5	8.3869 E−15	7.6724 E−15	5.6934 E−13
Butterfly	2	1.4179 E−15	8.4310 E−09	7.5611 E−12
	3	3.0199 E−08	1.2170 E−04	9.6050 E−08
	4	3.7441 E−11	5.0935 E−12	8.2234 E−13
	5	5.1381 E−08	7.3796 E−15	4.8668 E−09
Maize	2	7.9676 E−11	7.1349 E−16	3.2984 E−08
	3	9.0006 E−11	2.9541 E−06	4.6093 E−11
	4	9.0030 E−07	6.9312 E−04	6.8892 E−15
	5	1.5321 E−14	9.3836 E−13	8.2699 E−04
Sea star	2	1.8347 E−15	9.2729 E−15	9.6341 E−06
	3	2.1182 E−13	1.1408 E−12	9.6717 E−16
	4	3.2643 E−07	2.5590 E−14	3.9884 E−16
	5	7.6816 E−16	8.6944 E−12	6.4834 E−04
Smiling girl	2	3.1091 E−14	9.1850 E−08	7.9916 E−06
	3	3.3765 E−16	3.8180 E−06	8,8123 E−08
	4	7.3174 E−11	6.2570 E−07	4.1653 E−14
	5	8.5530 E−09	7.9818 E−08	2.7146 E−12
Surfer	2	3.4667 E−08	5.0517 E−16	9.7685 E−13
	3	7.3319 E−14	1.1479 E−13	2.5258 E−15
	4	8.8110 E−13	3.1081 E−14	3.3225 E−15
	5	2.6798 E−11	6.4653 E−09	3.5506 E−17
Train	2	3.0442 E−13	7.9150 E−17	7.4060 E−09
	3	4.6265 E−12	8.2253 E−16	7.6292 E−12
	4	5.5065 E−12	6.8620 E−17	1.6333 E−09
	5	8.1792 E−07	9.5124 E−13	4.6672 E−07

Table 4.17 Result after applying the MTEMO to the set of benchmark images

Image	k	Thresholds x_t^B	PSNR	STD	SSIM	FSIM
Camera man	2	71, 130	23.1227	31.00 E−04	0.9174	0.8901
	3	71, 130, 193	18.0122	72.01 E−04	0.8875	0.8456
	4	44, 84, 120, 156	24.9589	86.01 E−03	0.9363	0.9149
	5	44, 84, 120, 156, 196	23.0283	7.90 E−01	0.9289	0.8960
Lena	2	79, 127	23.9756	7.21 E−05	0.9083	0.8961
	3	79, 127, 177	21.0043	14.37 E−04	0.8660	0.8197
	4	62, 94, 127, 161	24.0020	18.69 E−03	0.9057	0.8851
	5	62, 94, 127, 161, 194	23.3736	39.82 E−02	0.8956	0.8684
Baboon	2	15, 105	23.5906	18.51 E−06	0.9480	0.9437
	3	51, 105, 158	19.9394	28.78 E−02	0.9011	0.9059
	4	33, 70, 107, 143	23.5022	22.65 E−02	0.9530	0.9594
	5	33, 70, 107, 143, 179	21.9540	37.13 E−01	0.9401	0.9417
Hunter	2	60, 119	22.8774	17.89 E−04	0.9192	0.8916
	3	60, 119, 179	20.2426	54.12 E−04	0.9031	0.8652
	4	46, 90, 134, 178	22.4723	1.94 E−02	0.9347	0.9159
	5	46, 90, 134, 178, 219	22.4025	1.23 E−01	0.9349	0.9173
Airplane	2	69, 125	25.4874	17.31 E−04	0.9685	0.9239
	3	69, 125, 180	22.9974	17.89 E−04	0.9433	0.8909
	4	55, 88, 122, 155	28.5400	19.21 E−03	0.9848	0.9677
	5	55, 88, 122, 155, 188	26.4997	35.08 E−03	0.9663	0.9417
Peppers	2	70, 145	19.6654	54.83 E−02	0.8697	0.8378
	3	70, 145, 223	17.2736	1.31 E−01	0.8437	0.7534
	4	46, 88, 132, 175	21.8275	3.02 E−04	0.8976	0.8552
	5	46, 88, 132, 175, 223	21.1207	6.34 E−03	0.8976	0.8304
Living room	2	55, 111	22.6665	47.11 E−03	0.9116	0.8966
	3	55, 111, 179	18.0379	15.27 E−04	0.8482	0.8132
	4	42, 85, 124, 162	21.7235	93.35 E−03	0.9170	0.9090
	5	42, 85, 124, 162, 201	21.3118	94.32 E−03	0.9183	0.9029
Blonde	2	62, 110	25.8389	31.91 E−04	0.9645	0.9503
	3	62, 110, 155	21.5001	37.05 E−04	0.9012	0.8759
	4	36, 65, 100, 134	25.9787	17.45 E−03	0.9606	0.9491
	5	36, 65, 100, 134, 168	23.1835	48.20 E−03	0.9328	0.9077
Bridge	2	65, 131	20.1408	22.71 E−04	0.8619	0.8749
	3	65, 131, 191	18.7016	40.49 E−04	0.8410	0.8479
	4	45, 88, 131, 171	21.4247	38.48 E−03	0.9168	0.9279
	5	45, 88, 131, 171, 211	21.0157	66.16 E−03	0.9153	0.9217
Butterfly	2	83, 120	26.7319	96.11 E−03	0.9493	0.9195
	3	83, 120, 156	24.4582	39.04 E−03	0.9386	0.8934
	4	70, 94, 119, 144	27.0221	14.59 E−02	0.9653	0.9417
	5	70, 94, 119, 144, 172	25.7809	98.61 E−02	0.9610	0.9283
Lake	2	71, 121	27.8565	10.69 E−04	0.9729	0.9638
	3	71, 121, 173	23.7695	12.87 E−04	0.9399	0.9288
	4	41, 80, 119, 159	24.7454	11.97 E−03	0.9587	0.9422
	5	41, 80, 119, 159, 197	22.4347	11.80 E−03	0.9439	0.9213

Table 4.18 Results after applying the MT-EMO using Tsallis entropy over the selected benchmark images

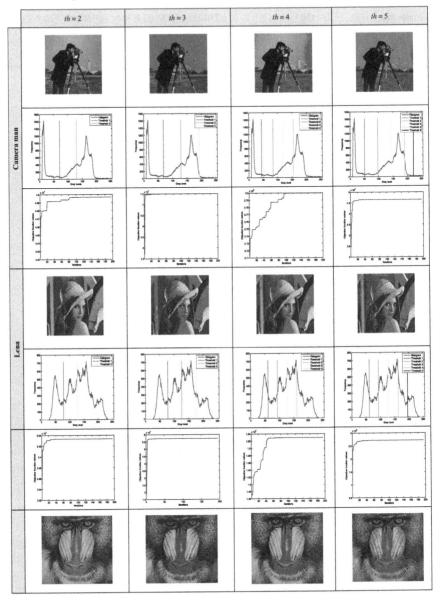

Comparisons of Tsallis entropy

In order to demonstrate that the TSEMO is an interesting alternative for MT, the proposed algorithm is compared with two state-of-the-art implementations. The methods used for comparison are: the Cuckoo Search Algorithm (CSA) [49] and the

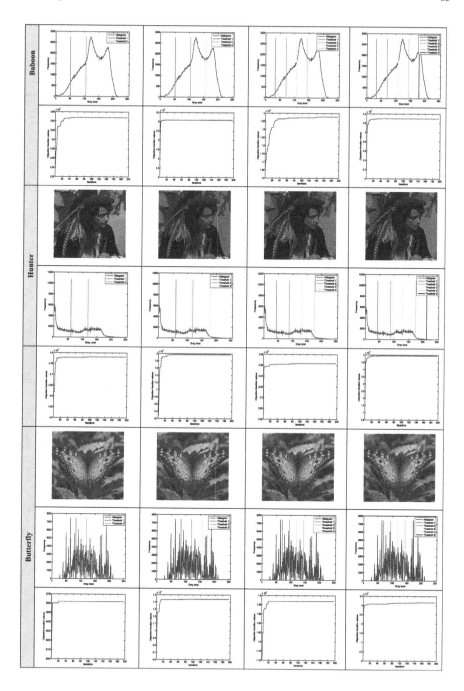

Table 4.19 Comparison of the *STD* and mean values of the TSEMO, CSA and PSO applied over the selected test images using Tsallis method

Image	k	TSEMO		CSA		PSO	
		STD	Mean	STD	Mean	STD	Mean
Camera man	2	31.00 E−04	4.49 E+04	89.56 E−04	4.02 E+04	83.00 E02	4.19 E+04
	3	72.01 E−04	7.49 E+04	98.32 E−04	6.99 E+04	89.00 E+00	7.27 E+04
	4	86.01 E−03	2.79 E+06	18.68 E−03	2.18 E+06	12.35 E+02	2.37 E+06
	5	7.90 E−01	4.65 E+06	69.98 E−01	4.56 E+06	5.38 E+03	4.28 E+06
Lena	2	7.21 E−05	3.43 E+04	2.61 E+00	3.33 E+04	15.27 E+00	3.30 E+04
	3	14.37 E−04	5.72 E+04	3.39 E+00	5.67 E+04	3.31 E+00	5.62 E+04
	4	18.69 E−03	1.62 E+06	5.52 E+00	1.45 E+06	7.35 E+00	1.45 E+06
	5	39.82 E−02	2.71 E+06	8.50 E+01	2.55 E+06	2.92 E+00	2.59 E+06
Baboon	2	18.51 E−06	3.64 E+04	15.11 E−02	3.47 E+04	2.64 E+00	3.40 E+04
	3	28.78 E−02	6.08 E+04	40.80 E−02	6.05 E+04	1.44 E+00	6.03 E+04
	4	22.65 E−02	1.97 E+06	62.02 E−02	1.90 E+06	8.11 E+00	1.86 E+06
	5	37.13 E−01	3.29 E+06	52.74 E−02	3.20 E+06	2.68 E+00	3.20 E+06
Hunter	2	17.89 E−04	4.78 E+04	7.38 E−04	4.70 E+04	4.38 E+00	4.72 E+04
	3	54.12 E−04	7.97 E+04	2.95E−04	7.89 E+04	9.47 E+00	7.85 E+04
	4	1.94 E−02	2.96 E+06	1.62 E−01	2.93 E+06	1.04 E+01	2.92 E+04
	5	1.23 E−01	4.94 E+06	2.46 E−01	4.89 E+06	3.23 E+02	4.75 E+04
Butterfly	2	96.11 E−03	8.61 E+03	12.78 E−02	8.56 E+03	6.36 E−01	8.55 E+03
	3	39.04 E−03	1.43 E+04	19.00 E−02	1.38 E+04	11.56 E−01	1.35 E+04
	4	14.59 E−02	1.88 E+05	11.04 E−01	1.80 E+05	1.04 E+00	1.81 E+05
	5	98.61 E−02	3.14 E+05	1.58 E+00	3.07 E+05	3.58 E+00	2.96 E+05

Table 4.20 Comparison of the *PSNR*, *SSIM* and *FSIM* values of the TSEMO, CSA and PSO applied over the selected test images using Tsallis method

Image	k	TSEMO			CSA			PSO		
		PSNR	SSIM	FSM	PSNR	SSIM	FSIM	PSNR	SSIM	FSIM
Camera man	2	23.1227	0.9174	0.8901	23.1194	0.9173	0.8901	22.9737	0.9160	0.8871
	3	18.0998	0.8875	0.8509	18.7480	0.8918	0.8456	18.0122	0.8874	0.8441
	4	25.0021	0.9369	0.9151	24.5479	0.9349	0.9097	23.3230	0.9280	0.8976
	5	22.9136	0.9286	0.8950	22.5284	0.9243	0.8891	21.9598	0.9222	0.8839
Lena	2	23.9982	0.9088	0.8966	23.9756	0.9083	0.8961	23.9594	0.9085	0.8953
	3	21.2592	0.8699	0.8255	20.9669	0.8655	0.8192	20.9989	0.8659	0.8196
	4	23.9783	0.9056	0.8849	23.9493	0.9056	0.8846	23.8175	0.9032	0.8815
	5	23.4275	0.8954	0.8691	23.3099	0.8960	0.8689	23.3777	0.8949	0.8674
Baboon	2	23.7510	0.9496	0.9452	23.5906	0.9480	0.9410	23.5048	0.9475	0.9323
	3	19.9386	0.9007	0.9057	19.9031	0.8810	0.8759	19.8021	0.8729	0.8729
	4	23.5165	0.9532	0.9593	23.5106	0.9270	0.9295	23.5163	0.9125	0.9159
	5	22.0538	0.9410	0.9408	21.9071	0.9399	0.9112	21.7165	0.9350	0.9377
Hunter	2	22.8783	0.9192	0.8916	22.8074	0.9089	0.8826	22.7910	0.9093	0.8818
	3	20.2581	0.9034	0.8654	20.0026	0.8931	0.8552	20.0858	0.8921	0.8521
	4	22.4221	0.9341	0.9159	21.3972	0.9237	0.9055	21.5061	0.9244	0.9024
	5	22.5014	0.9355	0.9199	21.3171	0.9236	0.9063	21.3754	0.9254	0.9005
Butterfly	2	26.8352	0.9504	0.9212	25.7319	0.9493	0.9195	25.1635	0.9431	0.9150
	3	24.4144	0.9383	0.8926	23.4545	0.9300	0.8834	23.5251	0.9315	0.8846
	4	27.1226	0.9653	0.9420	26.0314	0.9653	0.9317	26.0810	0.9653	0.9321
	5	25.8838	0.9609	0.9285	24.0086	0.9516	0.9201	24.4870	0.9533	0.9142

Particle Swarm Optimization (PSO) [19], both methods uses the Tsallis entropy. Similar to the previous test, all the algorithms run 35 times over each selected image. The images used for this test are Camera man, Lena, Baboon, Hunter and Butterfly. For each image is computed the *PSNR*, *STD*, *SSIM*, *FSIM* values and the mean of the objective function.

The comparison results between the three methods are divided in two tables, Table 4.19 shows the *STD* and mean values of the fitness function. Table 4.20 presents the values of the quality metrics obtained after apply the thresholds over the test images.

The fitness values of four methods are statistically compared using a non-parametric significance proof known as the Wilcoxon's rank test [47, 48] that is conducted with 35 independent samples. Such proof allows assessing result differences among two related methods. The analysis is performed considering a 5 % significance level over the best fitness (Tsallis entropy) value data corresponding to the five threshold points. Table 4.21 reports the p-values produced by Wilcoxon's test for a pair-wise comparison of the fitness function between two groups formed as TSEMO versus CSA, TSEMO versus PSO. As a null hypothesis, it is assumed that there is no difference between the values of the two algorithms

Table 4.21 Wilcoxon
p-values of the compared
algorithm TSEMO versus
CSA and TSEMO versus PSO

Image	k	p-values	
		TSEMO versus CSA	TSEMO versus PSO
Camera man	2	6.2137 E−07	8.3280 E−06
	3	1.0162 E−07	2.0000 E−03
	4	8.8834 E−08	13.710 E−03
	5	16.600 E−03	50.600 E−03
Lena	2	3.7419 E−08	1.6604 E−04
	3	1.4606 E−06	1.3600 E−02
	4	1.2832 E−07	2.9000 E−03
	5	3.9866 E−05	8.9000 E−03
Baboon	2	1.5047 E−06	2.5500 E−02
	3	6.2792 E−05	5.1000 E−03
	4	2.1444 E−12	3.3134 E−05
	5	2.1693 E−11	1.8000 E−03
Hunter	2	2.2100 E−02	2.2740 E−02
	3	3.6961 E−04	1.1500 E−02
	4	6.8180 E−02	9.9410 E−09
	5	5.8200 E−02	2.4939 E−04
Airplane	2	3.0000 E−03	6.6300 E−03
	3	7.6000 E−03	3.5940 E−02
	4	4.8092 E−12	1.1446 E−06
	5	1.0023 E−09	2.7440 E−02
Peppers	2	2.7419 E−04	1.3194 E−04
	3	2.6975 E−08	3.5380 E−02
	4	1.5260 E−08	6.0360 E−02
	5	7.2818 E−08	7.6730 E−02
Living room	2	1.4000 E−03	2.6340 E−02
	3	6.8066 E−08	2.8000 E−03
	4	8.7456 E−07	5.8730 E−03
	5	1.7000 E−03	5.1580 E−03
Blonde	2	3.0000 E−03	4.1320 E−02
	3	5.9000 E−03	8.9300 E−02
	4	1.3800 E−02	2.7700 E−02
	5	2.3440 E−02	5.6000 E−03
Bridge	2	1.5000 E−03	1.5700 E−02
	3	1.4300 E−02	1.5350 E−02
	4	1.7871 E−06	7.0400 E−03
	5	8.7000 E−03	1.2400 E−02

(continued)

Table 4.21 (continued)

Image	k	p-values	
		TSEMO versus CSA	TSEMO versus PSO
Butterfly	2	1.5000 E−03	1.1150 E−02
	3	3.1800 E−02	1.3760 E−02
	4	4.8445 E−07	8.1800 E−03
	5	1.6000 E−02	1.0630 E−02
Lake	2	7.6118 E−06	2.9500 E−02
	3	1.2514 E−06	6.5644 E−06
	4	2.2366 E−10	6.6000 E−03
	5	5.3980 E−06	9.4790 E−03

Table 4.22 Fitness comparison of PSO (blue line), CSA (black line) and EMO (red line) applied for multilevel thresholding using TE

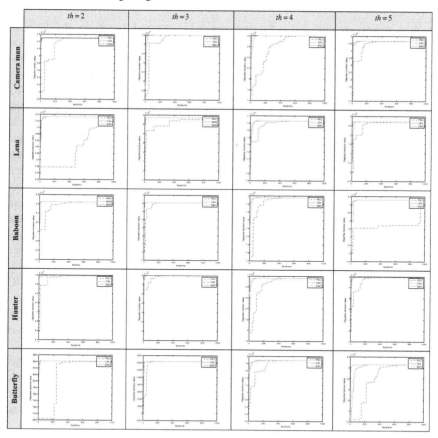

tested. The alternative hypothesis considers an existent difference between the values of both approaches. All p-values reported in Table 4.21 are less than 0.05 (5 % significance level) which is a strong evidence against the null hypothesis, indicating that the TSEMO fitness values for the performance are statistically better and it has not occurred by chance.

On the other hand, to compare the fitness of the three methods Table 4.21 shows the fitness values obtained for the reduced set of image (5 images). Each algorithm runs 1000 times and the best value of each run is stored, at the end of the evolution process the best stored values are plotted. From Table 4.22 it is possible to analyse that TSEMO and CSA reach the maximum entropy values in less iterations than the PSO method.

4.5 Conclusions

In this chapter, a multilevel thresholding (MT) method based on the Electro-magnetism-Like algorithm (EMO) is presented. The approach combines the good search capabilities of EMO algorithm with objective functions proposed by the popular MT methods of Otsu, Kapur and Tsallis. In order to measure the perfor-mance of the presented approach, it is used the peak signal-to-noise ratio (PSNR) which assesses the segmentation quality, considering the coincidences between the segmented and the original images.

The study explores the comparison between the two versions of MTEMO, one using the Otsu objective function and the other with the Kapur criterion. The results show that the Otsu function presents better results than the Kapur criterion. Such conclusion was statistically proved considering the Wilcoxon test.

The presented approach was compared with other techniques that implement different optimization algorithms like GA, PSO, CSA and BF. The efficiency of the algorithms was evaluated in terms of the PSNR and STD values. The experimental results provide evidence of the outstanding performance, accuracy and convergence of the presented algorithm in comparison with the other methods.

On the other hand for the Tsallis entropy implementation, thee fitness of TSEMO is compared with the CSA and PSO where is possible to see that both EMO and CSA need a reduced number of iterations to converge. However the speed of convergence of EMO is higher than de CSA in the same way th PSO is the slower and it has lack of accuracy.

Although the results offer evidence to demonstrate that the EMO method can yield good results on complicated images, the aim of our chapter is not to devise a multilevel thresholding algorithm that could beat all currently available methods, but to show that electro-magnetism systems can be effectively considered as an attractive alternative for this purpose.

References

1. Ghamisi, P., Couceiro, M.S., Benediktsson, J.A., Ferreira, N.M.F.: An efficient method for segmentation of images based on fractional calculus and natural selection. Expert Syst. Appl. **39**, 12407–12417 (2012)
2. Lai, C., Tseng, D.: A hybrid approach using Gaussian smoothing and genetic algorithm for multilevel thresholding. Int. J. Hybrid Intell. Syst. **1**, 143–152 (2004)
3. Hammouche, K., Diaf, M., Siarry, P.: A comparative study of various meta-heuristic techniques applied to the multilevel thresholding problem. Eng. Appl. Artif. Intell. **23**, 676–688 (2010)
4. Akay, B.: A study on particle swarm optimization and artificial bee colony algorithms for multilevel thresholding. Appl. Soft Comput. (2012). DOI 10.1016/j.asoc.2012.03.072
5. Liao, P.-S., Chung, P.-C., Chen, T.-S.: A fast algorithm for multilevel thresholding. J. Inf. Sci. Eng. **17**, 713–727 (2001)
6. Otsu, N.: A threshold selection method from gray-level histograms. In: IEEE Transactions on Systems, Man, Cybernetics, SMC-9, pp. 62–66 (1979)
7. Kapur, J.N., Sahoo, P.K., Wong, A.K.C.: A new method for gray-level picture thresholding using the entropy of the histogram. Comput. Vis. Graph. Image Process. **2**, 273–285 (1985)
8. Kittler, J., Illingworth, J.: Minimum error thresholding. Pattern Recogn. **19**, 41–47 (1986)
9. Sezgin, M., Sankur, B.: Survey over image thresholding techniques and quantitative performance evaluation. J. Electron. Imag. **13**, 146–168 (2004)
10. Sathya, P.D., Kayalvizhi, R.: Optimal multilevel thresholding using bacterial foraging algorithm. Expert Syst. Appl. **38**, 15549–15564 (2011)
11. Tsallis, C.: Possible generalization of Boltzmann-Gibbs statistics. J. Stat. Phys. **52**, 479–487 (1988)
12. Agrawal, S., Panda, R., Bhuyan, S., Panigrahi, B.K.: Tsallis entropy based optimal multilevel thresholding using cuckoo search algorithm. Swarm Evol. Comput. **11**, 16–30 (2013)
13. Portes de Albuquerque, M., Esquef, I.A., Gesualdi Mello, A.R.: Image thresholding using Tsallis entropy. Pattern Recogn. Lett. **25**(9), 1059–1065 (2004)
14. Tsallis, C., Abe, S., Okamoto, Y.: Non-extensive statistical mechanics and its applications. In: Series Lecture Notes in Physics. Springer, Berlin (2001)
15. Tsallis, C.: Entropic non extensivity: a possible measure of complexity. Chaos Solitons Fractals **13**, 371–391 (2002)
16. Rényi, A.: On Measures of Entropy and Information, pp. 547–561. University California Press, Berkeley (1988)
17. Goldberg, D.E.: Genetic Algorithms in Search, Optimization and Machine Learning, 1st edn. Addison-Wesley Longman Publishing Co. Inc., Boston (1989)
18. Yin, P.-Y.: A fast scheme for optimal thresholding using genetic algorithms. Sig. Process. **72**, 85–95 (1999)
19. Kennedy, J., Eberhart, R.C.: Particle swarm optimization. In: Proceedings of IEEE International Conference on Neural Networks, pp. 1942–1948. Piscataway (1995)
20. Horng, M.: Multilevel thresholding selection based on the artificial bee colony algorithm for image segmentation. Expert Syst. Appl. **38**, 13785–13791 (2011)
21. Das, S., Biswas, A., Dasgupta, S., Abraham, A.: Bacterial foraging optimization algorithm: theoretical foundations, analysis, and applications. Stud. Comput. Intell. **203**, 23–55 (2009)
22. Sathya, P.D., Kayalvizhi, R.: Optimal multilevel thresholding using bacterial foraging algorithm. Expert Syst. Appl. **38**, 15549–15564 (2011)
23. Sarkar, S., Das, S., Chaudhuri, S.: Multilevel image thresholding based on Tsallis entropy and differential evolution. In: Swarm, Evolutionary, and Memetic Computing, Lecture Notes in Computer Science, vol. 7677, pp. 17–24. Springer, Berlin (2012)

24. Storn, R., Price, K.: Differential evolution—a simple and efficient heuristic for global optimization over continuous spaces. J. Glob. Optim. **11**(4), 341–359 (1995)
25. Kayalvizhi, R., Sathya, P.D.: PSO-based Tsallis thresholding selection procedure for image segmentation. Int. J. Comput. Appl. **5**(4), 39–46 (2010)
26. Karaboga, D., Basturk, B.: A powerful and efficient algorithm for numerical function optimization: artificial bee colony (ABC) algorithm. J. Glob. Optim. **39**(3), 459–471 (2007)
27. Zhang, Y., Wu, L.: Optimal multi-level thresholding based on maximum Tsallis entropy via an artificial bee colony approach. Entropy **13**(4), 841–859 (2011)
28. Yang, X., Deb, S.: Cuckoo search via levy flights. In: Proceedings of the World Congress on Nature and Biologically Inspired Computing, vol. 4, pp. 210–214. NABIC, Coimbatore (2009)
29. Das, S., Biswas, A., Dasgupta, S., Abraham, A.: Bacterial foraging optimization algorithm: theoretical foundations, analysis, and applications. Stud. Comput. Intell. Found. Comput. Intell. **3**, 23–55 (2009)
30. Agrawal, S., Panda, R., Bhuyan, S., Panigrahi, B.K.: Tsallis entropy based optimal multilevel thresholding using cuckoo search algorithm. Swarm Evol. Comput. **11**, 16–30 (2013)
31. Sathya, P.D., Kayalvizhi, R.: Optimal multilevel thresholding using bacterial foraging algorithm. Expert Syst. Appl. **38**, 15549–15564 (2011)
32. Birbil, S.I., Fang, S.-C.: An electromagnetism-like mechanism for global optimization. J. Glob. Optim. **25**, 263–282 (2003)
33. Rocha, A., Fernandes, E.: Hybridizing the electromagnetism-like algorithm with descent search for solving engineering design problems. Int. J. Comput. Math. **86**, 1932–1946 (2009)
34. Rocha, A., Fernandes, E.: Modified movement force vector in an electromagnetism-like mechanism for global optimization. Optim. Meth. Softw. **24**, 253–270 (2009)
35. Tsou, C.S., Kao, C.H.: Multi-objective inventory control using electromagnetism-like metaheuristic. Int. J. Prod. Res. **46**, 3859–3874 (2008)
36. Wu, P., Yang, W.-H., Wei, N.-C.: An electromagnetism algorithm of neural network analysis an application to textile retail operation. J. Chin. Inst. Ind. Eng. **21**(1), 59–67 (2004)
37. Birbil, S.I., Fang, S.C., Sheu, R.L.: On the convergence of a population-based global optimization algorithm. J. Glob. Optim. **30**(2), 301–318 (2004)
38. Naderi, B., Tavakkoli-Moghaddam, R., Khalili, M.: Electromagnetism-like mechanism and simulated annealing algorithms for flowshop scheduling problems minimizing the total weighted tardiness and makespan. Knowl.-Based Syst. **23**, 77–85 (2010)
39. Hung, H.-L., Huang, Y.-F.: Peak to average power ratio reduction of multicarrier transmission systems using electromagnetism-like method. Int. J. Innovative Comput. Inf. Control **7**(5A), 2037–2050 (2011)
40. Yurtkuran, A., Emel, E.: A new hybrid electromagnetism-like algorithm for capacitated vehicle routing problems. Expert Syst. Appl. **37**, 3427–3433 (2010)
41. Jhang, J.-Y., Lee, K.-C.: Array pattern optimization using electromagnetism-like algorithm, AEU. Int. J. Electron. Commun. **63**, 491–496 (2009)
42. Lee, C.H., Chang, F.K.: Fractional-order PID controller optimization via improved electromagnetism-like algorithm. Expert Syst. Appl. **37**, 8871–8878 (2010)
43. Cuevas, E., Oliva, D., Zaldivar, D., Pérez-Cisneros, M., Sossa, H.: Circle detection using electromagnetism optimization. Inf. Sci. **182**(1), 40–55 (2012)
44. Guan, X., Dai, X., Li, J.: Revised electromagnetism-like mechanism for flow path design of unidirectional AGV systems. Int. J. Prod. Res. **49**(2), 401–429 (2011)
45. Hammouche, K., Diaf, M., Siarry, P.: A multilevel automatic thresholding method based on a genetic algorithm for a fast image segmentation. Comput. Vis. Image Underst. **109**, 163–175 (2008)
46. Pal, S.K., Bhandari, D., Kundu, M.K.: Genetic algorithms, for optimal image enhancement. Pattern Recogn. Lett. **15**, 261–271 (1994)
47. Wilcoxon, F.: Individual comparisons by ranking methods. Biometrics **1**, 80–83 (1945)

48. Garcia, S., Molina, D., Lozano, M., Herrera, F.: A study on the use of non-parametric tests for analyzing the evolutionary algorithms' behaviour: a case study on the CEC'2005 special session on real parameter optimization. J. Heurist. (2008). doi:10.1007/s10732-008-9080-4
49. Yang, X., Deb, S.: Engineering optimization by cuckoo search. Int. J. Math. Modell. Numer. Optim. **4**, 330–343 (2010)

Chapter 5
Template Matching Using a Physical Inspired Algorithm

5.1 Introduction

Locate and recognize objects in digital images are important fields of research in computer vision an image processing. These tasks are applied in many areas included industrial inspection, remote sensing, target classification and other important processes [1–3]. Template matching (TM) is an image processing technique to find object in images. In a TM approach, it is sought the point in which it is presented the best possible resemblance between a sub-image known as template and its coincident region within a source image.

In general TM involves two critical points: the similarity measurement and the search strategy [4]. Several metrics have been proposed to evaluate the matching between two images, the most important are: sum of absolute differences (SAD), sum of squared differences (SSD) and the normalized cross-correlation (NCC). The most used matching criterion is the NCC coefficient which is computationally expensive and represents the most consuming operation in the TM process [5].

The full search algorithm is the simplest TM algorithm that can deliver the optimal detection with respect to a maximal NCC coefficient as it checks all pixel-candidates one at a time. Unfortunately, such exhaustive search and the NCC calculation at each checking point, yields an extremely computational expensive TM method that seriously constraints its use for several image processing applications.

Recently, several TM algorithms, based on evolutionary approaches, have been proposed to reduce the number of NCC operations by calculating only a subset of search locations. Such approaches have produced several robust detectors using different optimization methods such as Genetic algorithms (GA) [6], Particle Swarm Optimization (PSO) [7, 8] and Imperialist competitive algorithm (ICA) [9]. Although these algorithms allow reducing the number of search locations, they do

© Springer International Publishing AG 2017
D. Oliva and E. Cuevas, *Advances and Applications of Optimised Algorithms in Image Processing*, Intelligent Systems Reference Library 117, DOI 10.1007/978-3-319-48550-8_5

not explore the whole region effectively and often suffers premature convergence which conducts to sub-optimal detections.

Electromagnetism-like algorithm (EMO) is a population-based evolutionary algorithm which was firstly introduced by Birbil and Fang [10] to solve optimization models using bounded variables. The algorithm imitates the attraction–repulsion mechanism between charged particles in an electromagnetic field. Each particle represents a solution and carries a certain amount of charge which is proportional to the solution quality (objective function). In turn, solutions are defined by position vectors which give real positions for particles within a multi-dimensional space. Moreover, objective function values of particles are calculated considering such position vectors. Each particle exerts repulsion or attraction forces over other population members; the resultant force acting over a particle is used to update its position. Clearly, the idea behind the EMO methodology is to move particles towards the optimum solution by exerting attraction or repulsion forces. Different to GA, PSO and ICA, EMO exhibits interesting search capabilities such as fast convergence whereas maintains the ability to avoid local minima [11–13].

EMO has been successfully applied to solve different sorts of engineering problems such as flow-shop scheduling [14], communications [15], vehicle routing [16], array pattern optimization in circuits [17], neural network training [18] control systems [19] and image processing [20]. Even though, the EMO algorithm has the capability to find the optimal values in complex optimization problems, it presents a critical problem in the Local Search (LS) stage [12, 13, 21]. Such stage is the most time consuming procedure of the approach, in which each particle-dimension is modified using a considerable number of iterations, in order to locally improve its quality.

One particular difficulty in applying an evolutionary algorithm, such as EMO, to discrete optimization problems is the multiple evaluation of the same individual. Discrete optimization problems are defined by using search spaces compound by a set of finite solutions. Therefore, since random numbers are involved in the calculation of new individuals, they may encounter the same solutions (repetition) that have been visited by other individuals at previous iterations, particularly when individuals are confined to a finite area. Evidently, such fact seriously constraints its performance mainly when fitness evaluation is computationally expensive to calculate.

In this chapter, an algorithm based on EMO is proposed to reduce the number of search locations in the TM process. The algorithm uses an enhanced EMO version where a modification of the local search procedure is incorporated in order to accelerate the exploitation process. Such modification reduces the number of perturbations around each particle to a compact number of random samples. As a result, the new EMO algorithm can substantially reduce the number of function evaluations yet preserving the good search capabilities of the original EMO. In the proposed approach, particles represent search positions which move throughout the positions of the source image. The NCC coefficient, used as a fitness value (charge extent), evaluates the matching quality presented between the template image and the coincident region of the source image, for a determined search position (particle). The number of NCC evaluations is also reduced by considering a memory which stores the NCC values previously visited in order to avoid re-evaluation of the same particles. Guided by the fitness values (NCC coefficients), the set of encoded candidate positions are evolved using the EMO operators until the best possible resemblance has been found. The proposed method achieves the best balance over other TM algorithms, in terms of both estimation accuracy and computational cost.

The remainder of this chapter is organized as follows: Sect. 5.2 explains a modification of the local search procedure in EMO. Section 5.3 introduces the implementation of the enhanced EMO algorithm for template matching. Section 5.4 presents the experimental results for the proposed approach over standard test images and some conclusions are drawn in Sect. 5.5.

5.2 Template Matching Process

Consider the problem of localizing a given reference image (template) defined as R within a larger intensity image know as I, which we call the source image. The task is to find those positions where the contents of the reference image R and the corresponding sub-image of I are either the same or most similar.

If it is denoted by $R_{u,v}(x,y) = R(x - u, y - v)$ where x and y is the position of R, the reference image R shifted by the distance (u,v) in the horizontal and vertical directions, respectively. Then the matching problem (illustrated in Fig. 5.1) can be summarized as follows: considering the source image I and the reference image R, find the offset (u,v) within the search region $\mathbf{S}$ such that the similarity between the shifted reference image $R_{u,v}(x,y)$ and the corresponding sub-image of I is a maximum.

In order to successfully solve this task, several issues need to be addressed such as determining a minimum similarity value to validate that a match has occurred and developing a good search strategy to find, in a fast way, the optimal displacement. Several TM algorithms that use evolutionary approaches [7–10] have been proposed as a search strategy to reduce the number of search positions. In comparison to other similarity criteria, NCC is the most effective and robust method that supports the measurement of the resemblance between R and its coincident region at image I, at each displacement (u,v). The NCC value between a given image I of size $M \times N$ and a template image R of size $m \times n$, at the displacement (u,v), is given by:

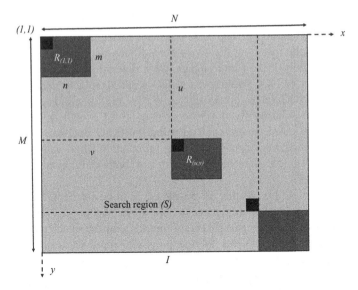

Fig. 5.1 Geometry of template matching. The reference image R is shifted across the search image I by an offset (u, v) using the origins of the two images as the reference points. The dimensions of the source image $(M \times N)$ and the reference image $(m \times n)$ determine the maximal search region (S) for this comparison

$$
NCC(u, v) = \frac{\sum\limits_{i=1}^{m} \sum\limits_{j=1}^{n} [I(u+i, v+j) - \bar{I}(u, v)] \cdot [R(i, j) - \bar{R}]}{\left[\sum\limits_{i=1}^{m} \sum\limits_{j=1}^{n} I(u+i, v+j) - \bar{I}(u, v) \right]^{\frac{1}{2}} \cdot \left[\sum\limits_{i=1}^{m} \sum\limits_{j=1}^{n} R(i, j) - \bar{R} \right]^{\frac{1}{2}}}
\tag{5.1}
$$

where $\bar{I}(u, v)$ is the grey-scale average intensity of the source-image for the coincident region of template image R whereas $\bar{R}$ is the grey-scale average intensity of the template image. These values are defined as follows:

$$
\bar{I}(u, v) = \frac{1}{m \cdot n} \sum\limits_{i=1}^{m} \sum\limits_{j=1}^{n} I(u+i, v+j) \, \bar{R} = \frac{1}{m \cdot n} \sum\limits_{i=1}^{m} \sum\limits_{j=1}^{m} R(i, j)
\tag{5.2}
$$

The NCC operation delivers values between the interval $[-1, 1]$, it means that if $NCC = 1$ the similarity is the best possible whereas if $NCC = -1$ the template and the corresponding image are completely different. Therefore, the point (u, v) which presents the best possible resemblance between R and I is defined as follows:

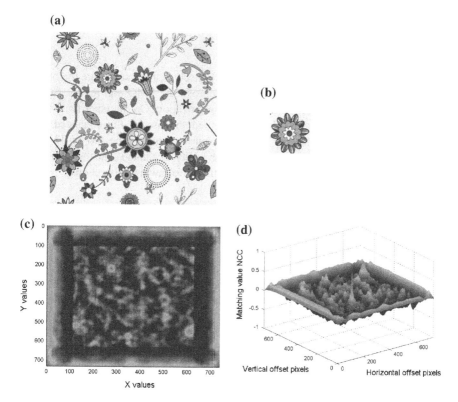

Fig. 5.2 Template matching process. **a** Example source image, **b** template image, **c** color-encoded NCC values and **d** NCC multi-modal surface

$$(u, v) = \arg \max_{(\hat{u}, \hat{v}) \in S} NCC(\hat{u}, \hat{v}) \tag{5.3}$$

where $S = \{(\hat{u}, \hat{v}) |\ 1 \leq \hat{u} \leq M - m, 1 \leq \hat{v} \leq N - n\}$.

Figure 5.2 illustrates the TM process considering Fig. 5.2a, b as the source and template image respectively. It is important to point out that the template image (5.2b) is similar but not equal to the coincident pattern, contained in the source image (5.2a). Figure 5.2c shows the NCC values (color-encoded) calculated in all locations of the search region S. On the other hand, Fig. 5.2d presents the NCC surface which exhibits the highly multi-modality nature of the TM problem. Basically Fig. 5.2c, d presents the search surface to show that it has several local maxima and only one global maxima, for that reason the classical search methods especially the gradient based could be trapped in the local optimal value.

5.3 New Local Search Procedure in EMO for Template Matching

In this section is introduced a new local search mechanism to substitute the original LS procedure in EMO. Such modification improve the EMO and reduce substantially the function calls required during the optimization. On the other hand, under the Template Matching approach are presented the use of the enhanced EMO algorithm and a Fitness Memory (FM). The purpose of use the FM in TM is to avoid the evaluation of points that were previously evaluated and their fitness is known. Such implementation are described in the following subsections.

5.3.1 The New Local Search Procedure for EMO

The local search procedure represents the exploitation phase of the EMO algorithm. Exploitation is the process of refining those points of a search space within the neighborhood of previously visited locations in order to improve their solution quality. In the literature [11, 22], it is proposed two approaches—local search applied to all points and local search applied only to the current best point. For the first, it has been proved in [11] that EMO presents the best possible properties of convergence and global optimization. However, under such circumstances, experimental studies [12, 22] show that the local search procedure spends more than the 80 % of the computational time for the complete EMO process. On the other hand, for the second, where the local search uses only the current best particle, experimental results provided in [22, 23] demonstrate that the EMO search capabilities are seriously weakened, mainly, when it is faced complex optimization functions.

In order to reduce the computational time without weakening the good search capabilities of the EMO algorithm, a new local search procedure is incorporated. The new procedure is a selective operation which is applied only to a subset $\mathbf{E}_t$ of the modified population $\mathbf{Y}_t$ (where $\mathbf{E}_t \subseteq \mathbf{Y}_t$). Under the new local search mechanism, it is created a subspace C_j around each selected particle $y_{j,t} \in \mathbf{E}_t$. Such subspace is generated by using an exploitation distance e_d defined as follows:

$$e_d = \frac{\sum_{q=1}^{n} \left(u_q - l_q \right)}{n} \cdot \beta \qquad (5.4)$$

where u_q and l_q are the upper and lower bounds in the q-th dimension, n is the number of dimensions of the optimization problem, whereas $\beta \in [0, 1]$ is a tuning factor. Therefore, the subspace bounds of C_j are modeled as follows:

$$uss_j^q = y_{j,t}^q + e_d$$
$$lss_j^q = y_{j,t}^q - e_d$$

(5.5)

where uss_j^q and lss_j^q are the upper and lower bounds of the q-th dimension for the subspace C_j, respectively.

Under the EMO context explained in Chap. 3, the LS generates set of perturbed particles defined as $Z_t = \{z_{1,t}, z_{2,t}, \ldots, z_{N,t}\}$. Considering the defined subspace C_j around each element of E_t, a set of m new particles ($P_j^m = \{p_j^1, p_j^2, \ldots, p_j^m\}$) are generated randomly inside the bounds defined by Eq. 5.5. Once the m samples are generated, the best sample p_j^{best} is selected in terms of the fitness values (where $p_j^{best} \in [p_j^1, p_j^2, \ldots, p_j^m]$) and its stored in G_t.

The new local search procedure is a selective operation which is applied only to a subset E_t of the modified population Y_t (where $E_t \subseteq Y_t$). The elements of E_t must fulfill two conditions. The first one is that each particle $y_{j,t} \in E_t$ must be part of the first half of the temporal population B ($B = \{b_1, b_2, \ldots, b_N\}$), produced after having been sorted Y_t according with the problem (descending for minimization and ascending for maximization) and with its fitness values. The second one is that inside C_j must not have located any other particle $y_{i,t}$ whose fitness values present a better extent than $y_{j,t}$. The first condition assures that only individuals with a high fitness value can be considered to be exploited. Since the EMO algorithm tends to concentrate to a determined solution as the method evolves, the number of particles to be exploited, according to the second condition, will be decremented iteration by iteration. Such behavior reflexes the fact that a high concentration of particles in a solution represents a kind of extensive exploitation itself. Under such conditions, it is not necessary to apply local search again. Finally a comparison between the values of B and E_t is performed in order to create Z_t the best values that exists between both sets are selected considering the worst and the best elements of the modified population Y_t. The complete new local search procedure is described by the Algorithm 5.1.

Algorithm 5.1 New Local Search	
1.	$\mathbf{B} \leftarrow Sort(\mathbf{Y}_t)$
2.	**For** $j = 1$ to $N/2$ **do**
3.	$E_{j,t} \leftarrow Obtain\_elements(\mathbf{B})$
4.	$C_j \leftarrow Define\_Subspace(E_{j,t})$
5.	$y_{c,t} \leftarrow Obtain\_the\_closest\_element\_to(E_{j,t})$
6.	Flag1= $y_{c,t}$ is inside C_j
7.	Flag2= $f(E_{j,t}) \geq f(y_{c,t})$
8.	**if** ((Not(Flag1)) or (Flag1 and Flag2)) **then**
9.	$\mathbf{P}_j^m \leftarrow Generate\_m\_samples(C_j)$
10.	$p_j^{best} \leftarrow Obtain\_best(\mathbf{P}_j^m)$
11.	$G_j \leftarrow p_j^{best}$
12.	**if** $f(G_j) < f(E_{j,t})$ **then**
13.	$E_{j,t} \leftarrow G_j$
14.	**end if**
15.	**end if**
16.	**end for**
17	**For** $j = 1$ to N **do**
18.	**if** $j < N/2$ **and** $f(E_{j,t}) < f(B_j)$ **then**
19.	$Z_j \leftarrow E_{j,t}$
20.	**else**
21.	$Z_j \leftarrow B_j$
22.	**end if**
23	**end for**

In steps 12 and 18 the sign in the fitness comparison depends on if is a maximization or minimization problem. With the incorporation of the new local search procedure, it is obtained a modified EMO method. As a result, the approach can substantially reduce the number of function evaluations preserving the good search capabilities of the EMO algorithm.

In order to demonstrate the new local search procedure operation, a numerical example has been set by applying the presented process to a simple function. Such function considers the interval of $-3 \leq$ dimension1, dimension2 ≤ 3 whereas the function possesses one global maxima of value 8.1 (such value is provided by the own function defection) at $(0, 1.6)$, notice that dimension1 and dimension2 correspond to the function axis (commonly x and y). As an example, it is assumed a modified population $\mathbf{Y}_t$ of six 2-dimensional members ($N = 6$). Figure 5.3a shows the initial configuration of the proposed example, the black points represents the half of the particles with the best fitness values (the first half of $\mathbf{B}$) whereas the gray points corresponds to remaining particles. From Fig. 5.3a, it can be seen that the new local search procedure is applied to all black particles ($y_{1,1}$, $y_{3,1}$ and $y_{5,1}$) generating, for each, 2 new random particles ($\mathbf{P}_1^2$, $\mathbf{P}_3^2$ and $\mathbf{P}_5^2$) inside of their corresponding subspaces (C_1, C_3 and C_5). Such operation is executed over $y_{1,1}$, $y_{3,1}$ and $y_{5,1}$ because they fulfil the two necessary conditions. Considering as example the

Fig. 5.3 Operation of the new local search procedure, **a** operation considering the initial modified population $\mathbf{Y}_1$, **b** operation considering the 15th modified population $\mathbf{Y}_{15}$ and **c** operation considering the 25th modified population $\mathbf{Y}_{25}$

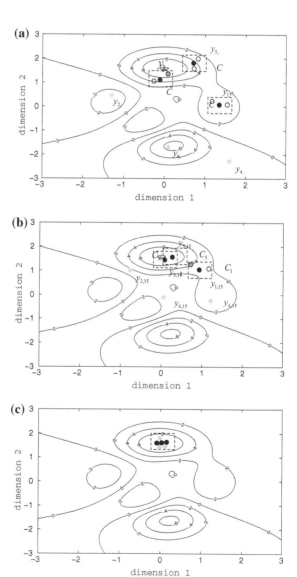

particle $y_{3,1}$ in Fig. 5.3a, the yellow particle p_3^{best}, randomly generated, will substitute the $y_{3,1}$ in the final individual $z_{3,2}$, since it possesses a better fitness value $(f(y_{3,1}) < f(p_3^{best}))$. Figure 5.3b shows the particle configuration after 14 new iterations. The particles which have been moved to new positions reduce the number of times that the new local search is executed. From Fig. 5.3b, it is evident that particles $y_{3,15}$ and $y_{5,15}$ are so close that they do not fulfil the necessary conditions to be processed by the new local search. In contrast, particle $y_{1,15}$ is processed generating two samples inside the subspace C_1. As the iteration number increases, the particles tend to concentrate around a solution. Figure 5.3c shows the particle configuration at the 25th iteration. As it can be seen, all particles are grouped around the optimal value. Under such circumstances, the new local search procedure is not executed, since such concentration works as a kind of exploitation process, where several particles try to refine those points of a search space within the neighborhood of a well-known solution.

5.3.2 Fitness Estimation for Velocity Enhancement in TM

The TM problem is defined by using a search space compound by a set of finite solutions. Therefore, since EMO algorithm employs random numbers for the calculation of new particles, such particles may encounter the same solutions (repetition) that have been visited by other individuals at previous iterations. Evidently, such fact seriously constraints the EMO performance mainly when fitness evaluation is computationally expensive to calculate, as it is in the case of the NCC computation.

In order to enhance the performance of the search strategy, the number of NCC evaluations is reduced by considering a fitness memory (**FM**) which stores the NCC values previously visited in order to avoid re-evaluation of the same particles. The **FM** memory contains a list that includes search position and its corresponding NCC value. Therefore, before evaluating a determined search position, it is analyzed the FM memory, if it already contains the search position, then it is not necessary to evaluate it; otherwise the NCC of the search position is computed and stored in the **FM** memory for its later use.

5.3.3 Template Matching Using EMO

In the EMO-based algorithm, individuals represent search positions (u, v) which move throughout the search space S. The NCC coefficient, used as a fitness value, evaluates the matching quality presented between the template image R and the source image I, for a determined search position (individual). The number of NCC evaluations is reduced by considering a memory which stores the NCC values previously visited in order to avoid re-evaluation of the same particles. Guided by the fitness values (NCC coefficients), the set of encoded candidate positions are evolved using the modified EMO operators until the best possible resemblance has been found.

In the algorithm, the search space S consists of a set of 2-D search positions $\hat{u}$ and $\hat{v}$ representing the components of the search location. Each particle is thus defined as:

$$x_i = \{(\hat{u}_i, \hat{v}_i) \mid 1 \leq \hat{u}_i \leq M - m, 1 \leq \hat{v}_i \leq N - n\} \tag{5.6}$$

The proposed EMO-based algorithm can be summarized in the following steps:

Step 1: Read gray scale image I.

Step 2: Select the template R.

Step 3: Initialize the **FM** memory in value outside the NCC range (for this chapter we use 2).

Step 4: Initialize randomly a set $\mathbf{X}_t$ of N particles each particle has $n = 2$ dimensions.

Step 5: Evaluate the entire population in the NCC function (Eqs. 5.3 and 5.4) and store the values in corresponding positions of **FM**.

Step 6: Compute the charge $q_{i,t}$ and the total force vector $F_{i,j}^t$ of the particles according with the Chap. 3.

Step 7: Move the particles to new positions based on the force.

Step 8: Before evaluate the NCC values for the modified particles, analyze the memory **FM**, in order to verify which positions have been already calculated.

Step 9: Apply the new local search procedure (Algorithm 5.1).

Step 10: Select the best particle x_t^B that has the higher NCC value.

Step 11: If the number of iterations has been reached, then determine the best individual (matching position) of the final population is $\hat{u}_{best}, \hat{v}_{best}$; otherwise go to step 6.

The proposed EMO-TM algorithm considers multiple search locations during the complete optimization process. However, only a few of them are evaluated using the true fitness function whereas all other remaining positions are just taken from the memory **FM**. Figure 5.4 shows a section of the search-pattern that has been generated by the EMO-TM approach considering the problem exposed in Fig. 5.3. Such pattern exhibits the evaluated search-locations in green-cells, whereas the maximum location is marked in red. Blue-cells represent those that have been repeated (they were multiply chosen) whereas cells with other gray intensity levels were not visited at all, during the optimization process.

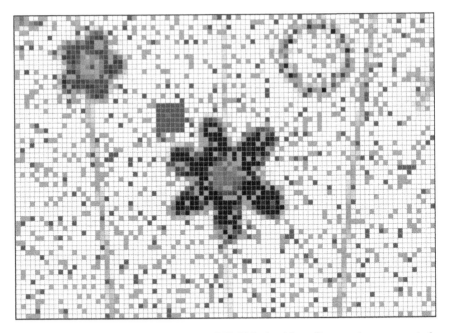

Fig. 5.4 Search-pattern generated by the EMO-TM algorithm. Green points represent the evaluated search positions whereas blue points indicate the repeated locations. The red point exhibits the optimal match detection

5.4   Experimental Results

In order to verify the feasibility and effectiveness of the proposed algorithm, a set of comparative experiments with other TM algorithms are also given. Simulations have been performed over a set of images which are shown in Table 5.1 with their respectively templates. To illustrate the complexity of TM as an optimization problem, Table 5.1 also presents the optimization surfaces (NCC surface) produced by the use of the full search approach. Since the NCC is illustrated, the values us such surfaces are between [−1, 1] and the global optimal is 1. The purpose of Table 5.1 is to show that the NCC surface are different for each image. No matter that the problem is to find the best NCC value, any surface is similar to other and they have several local optimal, for that reason traditional search algorithms fail on this task. On the other hand methods as full search needs to explore the entire image just to find the best NCC but if the image is too big they need to compute the fitness for each pixel. It means that if the best NCC is closer to the pixel position (1, 1) the full search algorithm explores the entire image and computes the NCC. Meanwhile the proposed optimization approach stops whit a stop criterion, it could be when the fitness doesn't change during a number of iterations or when a number of function calls or iterations is reached. In this way the EMO-TM with the use of FM reduces the number of NCC calculations.

Table 5.1 TMLSSEMO applied to different kind of images

Image		Template to search	NCC surface	Size (pixels)
Dog				574 × 800
Soccer				720 × 399
Waldo				1024 × 768
Airport				2048 × 1536
City				550 × 413
PCB				3220 × 2400

Then, the proposed approach has been applied to the experimental set, whose results have been compared to those produced by the ICA-TM method [9], standard Particle Swarm Optimization (PSO) [24], Differential Evolution (DE) [25] and the original EMO algorithm [10]. The first one is considered a state-of-the-art algorithm whose results have been recently published, PSO and DE are popular optimization methods that had been used in several applications. The maximum iteration number for the experimental set has been set to 300. Such stop criterion has been selected to maintain compatibility to similar works reported in the literature [6–9].

The parameter setting for each algorithm in the comparison is described as follows:

1. ICA-TM [10]: *NumOfCountries* = 100, *NumOfImper* = 10, *NumOf-Colony* = 90, $T_{max} = 300$, $\xi = 0.1$, $\varepsilon_1 = 0.15$ and $\varepsilon_2 = 0.9$. Such values are the best parameter set for this algorithm according to [10].
2. PSO [24]: *Swarm_size* = 50, *inertia_weight* = 0.3925, *particle_best_weight* = 2.55, *swarm_best_weight* = 1.33 and *Max_iter* = 150. Such values, according to [26] represent the best possible configuration.
3. DE [25]: *population_size* = 50, *crossover_probability* = 0.74, *differential_weight* = 0.93 and *Max_generation* = 150. Such values, according to [27] represent the best possible configuration.
4. The original EMO (O-EMO) [10, 11]: particle number = 50, $\delta = 0.001$, *LISTER* = 4 and *MaxIter* = 150. Such values, according to [10, 11] represent the best possible configuration.
5. EMO-TM: $N = 50$, $m = 5$, $\beta = 0.05$, and $Iter_{max} = 150$. After extensive experimentation, these values correspond to the best possible performance for the TM tasks.

Once all algorithms were configured with such values, they are used without modification during the experiments. The comparisons are analyzed considering three performance indexes: the average NCC value (ANcc), the success rate (Sr) and the Average number of search locations (AsL). The average NCC value (ANcc) indicates the average NCC value considering the total number of executions. The success rate (Sr) represents the number of executions in percentage in which the algorithms find out successfully the optimal detection point. The Average number of search locations (AsL) exhibits the number of checked locations which has been visited during a single experiment. Such performance index can be related with the average number of function evaluations that the NCC coefficient is computed. In order to assure statistic consistency, all these performance indexes are averaged considering a determined number of executions.

The results for 35 runs are reported in Table 5.2 where the best outcome for each image is boldfaced. According to this table, EMO-TM delivers better results than ICA, PSO, DE and O-EMO for all images. In particular, the test remarks the largest difference in the success rate (Sr) and the average number of search locations (AsL). Such facts are directly related to a better trade-off between convergence and computational overhead, and the incorporation of **FM** memory. Figure 5.5 present

Table 5.2 Performance comparison of ICA-TM, PSO, DE, O-EMO and the proposed approach for the experimental set shown in Table 5.1

Image	Algorithm	Average Ncc Value (ANcc)	Success rate (Sr) %	Average number of search locations (AsL)
Dog	ICA-TM	0.8856	70.21	29,500
	PSO	0.6681	34.28	35,680
	DE	0.5631	45.58	32,697
	O-EMO	0.8281	90.54	46,140
	EMO-TM	**1.0000**	**100**	**13,800**
Soccer	ICA-TM	0.5054	14.65	29,621
	PSO	0.4050	2.85	35,240
	DE	0.3753	22.98	30,987
	O-EMO	0.7150	20.34	45,000
	EMO-TM	**1.0000**	**100**	**16,920**
Waldo	ICA-TM	0.6587	60.43	28,786
	PSO	0.2154	2.05	32,169
	DE	0.2057	2156	31,875
	O-EMO	0.6422	70.33	44,775
	EMO-TM	**0.9598**	**98.00**	**16,044**
Airport	ICA-TM	0.6959	54.42	29,177
	PSO	0.5655	2.85	28,978
	DE	0.5676	28.56	25,921
	O-EMO	0.9170	59.51	45,580
	EMO-TM	**1.0000**	**100**	**17,650**
City	ICA-TM	0.2656	46.21	29,399
	PSO	0.1777	2.00	31,213
	DE	0.1583	20.56	30,578
	O-EMO	0.6166	74.75	44,651
	EMO-TM	**0.9843**	**97.00**	**15,830**
PCB	ICA-TM	0.3136	51.04	28,985
	PSO	0.2090	2.00	30,459
	DE	0.2015	18.26	36,987
	O-EMO	0.6921	74.81	47,689
	EMO-TM	**0.9067**	**92.65**	**16,990**

the matching evolution curve for each image considering the average best NCC value seen so-far for all the algorithms employed in the comparison. Such evolution graphs have been computed considering a single execution.

Table 5.2 demonstrates that EMO-TM present a better precision in reference to its counterparts. According to Table 5.2, the EMO-TM presents a better performance than the other algorithms in terms of effectiveness, since it detects practically in all experiments the optimal detection point. On the other hand, the proposed EMO-TM algorithm was able to reduce drastically the number of search locations.

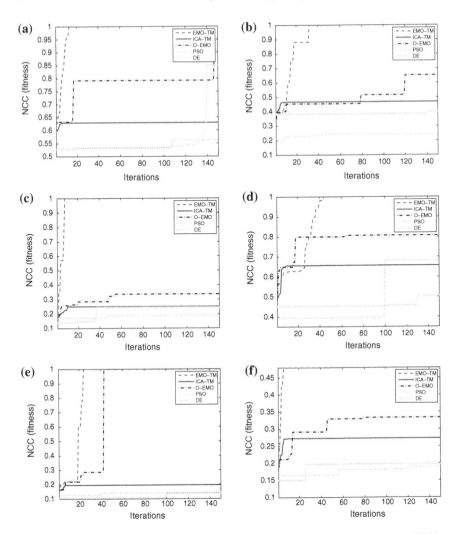

Fig. 5.5 Comparison of the NCC values for the images: **a** Dog, **b** Soccer game, **c** Waldo, **d** Airport, **e** City, **f** PCB, using the five different algorithms

Such value represents the number of NCC evaluations. It is important to recall that such evaluation represents the main computational cost associated to the TM process.

A non-parametric statistical significance proof known as the Wilcoxon's rank sum test for independent samples [28, 29] has been conducted over the average number of search locations (AsL) data of Table 5.2, with an 5 % significance level. Table 5.3 reports the p-values produced by Wilcoxon's test for the pair-wise comparison of the average number of search locations (AsL) of two groups. Such groups are formed by EMO-TM versus ICA-TM, EMO-TM versus PSO, EMO-TM

Table 5.3 *p*-values produced by Wilcoxon's test comparing EMO-TM versus ICA-TM, EMO-TM versus PSO, EMO-TM versus DE, and EMO-TM versus O-EMO over the average number of search locations (AsL) values from Table 5.2

Image	EMO-TM versus ICA-TM	EMO-TM versus PSO	EMO-TM versus DE	EMO-TM versus O-EMO
Dog	6.5304e−13	6.4894e−08	7.0548e−13	6.5154e−13
Soccer	6.5335e−13	8.5675e−11	2.9973e−12	6.5094e−13
Waldo	6.5304e−13	5.7977e−12	8.3789e−13	6.5034e−13
Airport	6.5304e−13	4.2026e−12	9.5761e−12	6.5064e−13
City	6.5154e−13	1.6791e−11	7.7189e−13	6.4825e−13
PCB	6.5304e−13	8.5921e−12	4.3589e−12	6.4462e−13

versus DE and EMO-TM versus O-EMO. As a null hypothesis, it is assumed that there is no significant difference between mean values of the two algorithms. The alternative hypothesis considers a significant difference between the AfE values of both approaches. All *p*-values reported in Table 5.3 are less than 0.05 (5 % significance level) which is a strong evidence against the null hypothesis. Therefore, such evidence indicates that EMO-TM results are statistically significant and that it has not occurred by coincidence (i.e. due to common noise contained in the process).

5.5 Conclusions

In this chapter, a new algorithm based on the Electromagnetism-Like algorithm (EMO) is presented to reduce the number of search locations in the TM process. The algorithm uses an enhanced EMO version where a modification of the local search procedure is incorporated in order to accelerate the exploitation process. Such modification reduces the number of perturbations around each particle to a compact number of random samples. As a result, the new EMO algorithm can substantially reduce the number of function evaluations yet preserving the good search capabilities of the original EMO.

In the presented approach, particles represent search positions which move throughout the positions of the source image. The NCC coefficient, used as a fitness value (charge extent), evaluates the matching quality presented between the template image and the coincident region of the source image, for a determined search position (particle). The number of NCC evaluations is also reduced by considering a memory which stores the NCC values previously visited in order to avoid re-evaluation of the same particles. Guided by the fitness values (NCC coefficients), the set of encoded candidate positions are evolved using the EMO operators until the best possible resemblance has been found. The proposed method achieves the best balance over other TM algorithms, in terms of both estimation accuracy and computational cost.

The performance of the approach based on EMO has been compared to other existing TM algorithms by considering different images which present a great variety of formats and complexities. Experimental results demonstrate the high performance of the proposed method in terms of precision and the number of NCC evaluations.

References

1. Brunelli, R.: Template Matching Techniques in Computer Vision: Theory and Practice. Wiley, New York (2009)
2. Crispin, A.J., Rankov, V.: Automated inspection of PCB components using a genetic algorithm template-matching approach. Int. J. Adv. Manuf. Technol. **35**, 293–300 (2007)
3. Juan, L., Jingfeng, Y., Chaofeng, G.: Research and implementation of image correlation matching based on evolutionary algorithm. In: Future Computer Science and Education (ICFCSE). 2011 International Conference, pp. 499, 501. 20–21 Aug 2011
4. Hadi, G., Mojtaba, L., Hadi, S.Y.: An improved pattern matching technique for lossy/lossless compression of binary printed Farsi and Arabic textual images. Int. J. Intell. Comput. Cybernet. **2**(1), 120–147 (2009)
5. Krattenthaler, W., Mayer, K.J., Zeiler, M.: Point correlation: a reduced-cost template matching technique. In: Proceedings of the First IEEE International Conference on Image Processing, pp. 208–212 (1994)
6. Dong, N., Wu, C.-H., Ip, W.-H., Chen, Z.-Q., Chan, C.-Y., Yung, K.-L.: An improved species based genetic algorithm and its application in multiple template matching for embroidered pattern inspection. Expert Syst. Appl. **38**, 15172–15182 (2011)
7. Liu, F., Duana, H., Deng, Y.: A chaotic quantum-behaved particle swarm optimization based on lateral inhibition for image matching. Optik **123**, 1955–1960 (2012)
8. Wu, C.-H., Wang, D.-Z., Ip, A., Wang, D.-W., Chan, C.-Y., Wang, H.-F.: A particle swarm optimization approach for components placement inspection on printed circuit boards. J. Intell. Manuf. **20**, 535–549 (2009)
9. Duan, H., Chunfang, X., Liu, S., Shao, S.: Template matching using chaotic imperialist competitive algorithm. Pattern Recogn. Lett. **31**, 1868–1875 (2010)
10. Ilker, B., Birbil, S., Shu-Cherng, F.: An electromagnetism-like mechanism for global optimization. J. Global Optim. **25**, 263–282 (2003)
11. Birbil, S.I., Fang, S.C., Sheu, R.L.: On the convergence of a population-based global optimization algorithm. J. Global Optim. **30**(2), 301–318 (2004)
12. Rocha, A., Fernandes, E.: Hybridizing the electromagnetism-like algorithm with descent search for solving engineering design problems. Int. J. Comput. Math. **86**, 1932–1946 (2009)
13. Rocha, A., Fernandes, E.: Modified movement force vector in an electromagnetism-like mechanism for global optimization. Optim. Methods Softw. **24**, 253–270 (2009)
14. Naderi, B., Tavakkoli-Moghaddam, R., Khalili, M.: Electromagnetism-like mechanism and simulated annealing algorithms for flowshop scheduling problems minimizing the total weighted tardiness and makespan. Knowl.-Based Syst. **23**, 77–85 (2010)
15. Hung, H.-L., Huang, Y.-F.: Peak to average power ratio reduction of multicarrier transmission systems using electromagnetism-like method. Int. J. Innovative Comput. **7**(5), 2037–2050 (2011)
16. Yurtkuran, A., Emel, E.: A new hybrid electromagnetism-like algorithm for capacitated vehicle routing problems. Expert Syst. Appl. **37**, 3427–3433 (2010)
17. Jhen-Yan, J., Kun-Chou, L.: Array pattern optimization using electromagnetism-like algorithm. AEU Int. J. Electron. Commun. **63**, 491–496 (2009)

18. Wu, P., Wen-Hung, Y., Nai-Chieh, W.: An electromagnetism algorithm of neural network analysis an application to textile retail operation. J. Chin. Inst. Ind. Eng. **21**, 59–67 (2004)
19. Lee, C.H., Chang, F.K.: Fractional-order PID controller optimization via improved electromagnetism-like algorithm. Expert Syst. Appl. **37**, 8871–8878 (2010)
20. Cuevas, E., Oliva, D., Zaldivar, D., Pérez-Cisneros, M., Sossa, H.: Circle detection using electro-magnetism optimization. Inf. Sci. **182**(1), 40–55 (2012)
21. Guan, Xianping, Dai, Xianzhong, Li, Jun: Revised electromagnetism-like mechanism for flow path design of unidirectional AGV systems. Int. J. Prod. Res. **49**(2), 401–429 (2011)
22. Lee, C.H., Chang, F.K.: Fractional-order PID controller optimization via improved electromagnetism-like algorithm. Expert Syst. Appl. **37**, 8871–8878 (2010)
23. Zhang, C., Li, X., Gao, L., Wu, Q.: An improved electromagnetism-like mechanism algorithm for constrained optimization. In: Expert Systems with Applications. doi:10.1016/j.eswa.2013. 04.028 (in Press)
24. Kennedy, J., Eberhart, R.C.: Particle swarm optimization. In: Proceedings of IEEE International Conference on Neural Networks, Piscataway. pp. 1942–1948 (1995)
25. Price, K., Storn, R.M., Lampinen, J.A.: Differential Evolution: A Practical Approach to Global Optimization. Springer, New York (2005)
26. Pedersen, M.E.H.: Good parameters for Particle Swarm Optimization. Technical report HL1001, Hvass Laboratories (2010)
27. Pedersen, M.E.H.: Good parameters for Differential Evolution. Technical report HL1002, Hvass Laboratories (2010)
28. Wilcoxon, F.: Individual comparisons by ranking methods. Biometrics **1**, 80–83 (1945)
29. Garcia, S., Molina, D., Lozano, M., Herrera, F.: A study on the use of non-parametric tests for analyzing the evolutionary algorithms' behaviour: a case study on the CEC'2005 Special session on real parameter optimization. J. Heurist. (2008). doi:10.1007/s10732-008-9080-4

Chapter 6
Detection of Circular Shapes in Digital Images

6.1 Introduction

Evolutionary computation [1] is a field of research that studies the use of biology as an inspiration to solve computational problems and the use of the natural world experiences to answer real-life challenges [2]. The increasing interest on the field emerges from the fact engineering is challenged by more complex, large, distributed and ill-structured systems. However, Nature is providing simple structures and organizations which are capable of dealing with most complex systems and tasks. Many nature-inspired approaches consider phenomena that exhibit some type of collective intelligence [3, 4], a concept that has been lately developed and applied to problems holding a complex behavioral pattern [5]. The approach follows the idea that a system is composed of decentralized individuals which can interact to other elements according to localized knowledge. Therefore, the overall image of the system emerges from aggregating individual interactions. Special kinds of artificial collective-individuals are the elements created by analogy with bees [6], ants [7] or charged particles [8]. By adding some features from optimization, pattern recognition, shape detection and machine learning, the nature-inspired algorithms with collective intelligence characteristics have recently gained considerable research interest from the computer vision community.

The problem of detecting circular shapes holds paramount importance for image analysis, in particular for industrial applications such as automatic inspection of manufactured products and components, aided vectorization of drawings, target detection, etc. [9, 10]. Solving object location challenges is normally approached by two types of techniques: Deterministic Techniques including the application of Hough transform based methods [11], geometric hashing and template or model matching techniques [12, 13] and Stochastic Techniques, which include random sample consensus techniques [14], simulated annealing [15] and Genetic Algorithms (GA) [16].

© Springer International Publishing AG 2017
D. Oliva and E. Cuevas, *Advances and Applications of Optimised Algorithms in Image Processing*, Intelligent Systems Reference Library 117,
DOI 10.1007/978-3-319-48550-8_6

Template and model matching techniques are the first approaches to be applied to shape detection yielding a considerable amount of publications [17]. Shape coding techniques and combination of shape properties have been commonly used to represent objects. The main drawback of such proposals is related to the contour extraction step from a real image which in turn exhibits great difficulty to deal with pose invariance, except for very simple objects.

Circle detection in digital images is commonly performed by the Circular Hough Transform [18]. A typical Hough-based approach employs an edge detector and uses edge information to infer locations and radius values. Peak detection is then performed by averaging, filtering and histogramming the transform space. However, such approach requires a large storage space given the required 3-D cells to cover all parameters (x, y, r) and a high computational complexity which yields a low processing speed. The accuracy of the extracted parameters for the detected circle is poor, particularly under the presence of noise [19]. For a digital image holding a significant width and height and a densely populated edge pixel map, the required processing time for Circular Hough Transform makes it prohibitive to be deployed in real time applications. In order to overcome such a problem, some other researchers have proposed new approaches based on the Hough transform, for instance the probabilistic Hough transform [20], the randomized Hough transform (RHT) [21] and the fuzzy Hough transform [22]. Alternative transformations have also been presented in the literature as the one proposed by Becker in [23]. Although those new approaches demonstrated faster processing speeds in comparison to the original Hough Transform, they still retain high noise sensitivity.

Stochastic search methods such as Genetic Algorithms (GA) are other alternatives to shape recognition in computer vision. In particular, GA has recently been used for important shape detection tasks following the works of Roth and Levine who have proposed the use of GA for primitive extraction of images [16] while Lutton et al. have added a further improvement of the aforementioned method [24] and Yao et al. have proposed a multi-population GA to detect ellipses [25]. In [26], GA has been used for template matching when the pattern has been the subject of an unknown affine transformation. Ayala–Ramirez et al., presented a GA based circle detector [27] which is capable of detecting multiple circles on real images albeit failing on detecting imperfect circular shapes. Another example is presented in [28] which discusses on soft computing techniques for shape classification. In the case of ellipsoidal detection, Rosin proposes an ellipse fitting algorithm that uses five points in [29] while in [30] Zhang and Rosin extends the later algorithm to fit data into super-ellipses. Most of such approaches perform circle detection under an acceptable computation time despite noisy conditions. However, they still fail at the time of facing complex conditions such as occlusion or superposition.

In this chapter, the circle detection task is considered as an optimization problem. Although there are several traditional optimization algorithms, they require substantial gradient information targeting the solution within a neighborhood of a given initial approximation. If the problem has more than one local solution, the convergence to the global solution may depend on the provided initial approximation [31].

On the other hand, nature-inspired methods with collective behaviour have been successfully applied for solving constrained global optimization problems [5, 32]. For instance, in [33] has been proposed the Electromagnetism-like optimization (EMO) algorithm which is a flexible and effective nature-based approach to search for the optimal solution of single objective optimization problems. EMO originates from the electromagnetism theory of physics by considering potential solutions as electrically charged particles spread around the solution space, where the charge of each particle depends on its objective function value. This algorithm utilizes a collective attraction-repulsion mechanism to move the particles towards optimality. This interaction mechanism among particles in EMO corresponds to the interaction among individuals produced by reproduction, crossover and mutation in Genetic Algorithms (GA) [34]. Likewise, the way in which EMO moves the particles around the search space corresponds to the Particle Swarm Optimization (PSO) [35] and to the Ant Colony Optimization (ACO) [36].

Although EMO algorithm shares some characteristics to other nature-inspired approaches, recent works (see [37–40]) have exhibited EMO's great savings on computation time and memory allocation surpassing other methods such as GA, PSO and ACO for finding a global optimum. As a result of EMO's economic computational cost and reassured convergence [41], it has been successfully applied to solve different sorts of engineering problems such as flow shop scheduling problems [42], vehicle routing [43], array pattern optimization in circuits [44] and neural network tanning [37]. However, to the best of our knowledge, EMO has not been yet applied to any computer-vision related task.

This chapter presents a circle detector method based on the EMO algorithm. Here, the detection process is considered to be similar to an optimization problem. The algorithm uses the encoding of three non-collinear edge points as candidate circles (x, y, r) in the edge-only image of the scene. Guided by the values of the objective function, the set of encoded candidate circles are evolved using the EMO algorithm so that they can fit into the actual circles on the edge map. The approach generates a sub-pixel circle detector which can effectively identify circles in actual images despite circular objects exhibiting significant occluded portions. Experimental evidence shows performance evidence for detecting circles under different conditions.

This chapter is organized as follows: Sect. 6.2 the approach is formulated explaining the application of the EMO algorithm for circle detection. Section 6.3 discusses on the results after EMO application under different image conditions. Finally, Sect. 6.4 presents the conclusions.

6.2 Circle Detection Using EMO

At this work, circles are represented by the parameters of the well-known second degree equation shown by Eq. 6.1, considering three points [19] in the edge-only space of the image. A pre-processing stage requires marking the object's contour by applying a single-pixel edge detection method. In the scope of this paper, such task is accomplished by the classical Canny algorithm, storing the locations for each edge point. All such points are the only potential candidates to define circles by considering triplets. They are stored within a vector array $\mathbf{E} = \{\mathbf{e}^1, \mathbf{e}^2, \ldots, \mathbf{e}^{Np}\}$ with Np being the total number of edge pixels contained in the image. In turn, the algorithm stores the (x_v, y_v) coordinates of each edge pixel belonging to the edge vector $\mathbf{e}^v$ ($e_1^v = x_v, e_2^v = y_v$).

In order to construct each circle candidate (or charged particles within the EMO framework), the indexes v_1, v_2 and v_3 of three non-collinear edge points must be combined, assuming that the circle's contour goes through points $\mathbf{e}_{v_1}$; $\mathbf{e}_{v_2}$; $\mathbf{e}_{v_3}$. A number of candidate solutions are generated randomly for the initial pool of particles. The solutions will thus evolve through the application of the EMO algorithm as the evolution takes place over the pool of particles until a minimum is reached and the best particle is considered as the solution. As the overall evolution process evolves, the objective function improves at each generation by discriminating non-plausible circles and locating others by avoiding visits to other image points. The discussion that follows clearly explains the required steps to formulate the circle detection task within the EMO framework.

6.2.1 Particle Representation

Each pool particle $\mathbf{C}$ considers three edge points which are stored following a position-related index within the edge array $\mathbf{E}$. In turn, the procedure will encode a particle as the circle that passes through three points $\mathbf{e}_i$, $\mathbf{e}_j$ and $\mathbf{e}_k$ ($\mathbf{C}^p = \{\mathbf{e}_i, \mathbf{e}_j, \mathbf{e}_k\}$). Each circle $\mathbf{C}$ is thus represented by three parameters: x_0, y_0 and r, being (x_0, y_0) the coordinates of the centre of the circle and r its radius. The equation of the circle passing through the three edge points can be computed as follows:

$$(x - x_0)^2 + (y - y_0)^2 = r^2 \tag{6.1}$$

considering

$$\mathbf{A} = \begin{bmatrix} x_j^2 + y_j^2 - (x_i^2 + y_i^2) & 2 \cdot (y_j - y_i) \\ x_k^2 + y_k^2 - (x_i^2 + y_i^2) & 2 \cdot (y_k - y_i) \end{bmatrix}$$

$$\mathbf{B} = \begin{bmatrix} 2 \cdot (x_j - x_i) & x_j^2 + y_j^2 - (x_i^2 + y_i^2) \\ 2 \cdot (x_k - x_i) & x_k^2 + y_k^2 - (x_i^2 + y_i^2) \end{bmatrix}, \tag{6.2}$$

$$x_0 = \frac{\det(\mathbf{A})}{4((x_j - x_i)(y_k - y_i) - (x_k - x_i)(y_j - y_i))},$$

$$y_0 = \frac{\det(\mathbf{B})}{4((x_j - x_i)(y_k - y_i) - (x_k - x_i)(y_j - y_i))}, \tag{6.3}$$

Fig. 6.1 Circle candidate (charged-particle) built from the combination of points p_i, p_j and p_k

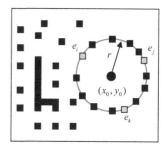

and

$$r = \sqrt{(x_0 - x_b)^2 + (y_0 - y_b)^2},\tag{6.4}$$

where det(.) stands for the determinant and $b \in \{i, j, k\}$. Figure 6.1 illustrates the parameters defined by Eqs. 6.2–6.4.

Therefore, it is possible to represent the shape parameters for each circle $[x_0, y_0, r]$ as a transformation T for all edge vector indexes i, j and k, yielding

$$[x_0, y_0, r] = T(i, j, k)\tag{6.5}$$

with T being the transformation calculated after the previous computations for x_0, y_0, and r.

By exploring each index as a particle, it is possible to sweep the continuous space by searching shape parameters by means of the EMO algorithm. This approach reduces the search space by eliminating unfeasible solutions.

6.2.2 Objective Function

A circumference may be calculated as a virtual shape as a way to measure the matching factor between the candidate circle **C** and the circle actually presented by the image. It must be validated if it really exists in the edge image. The test for these points is verified using $\mathbf{S} = \{\mathbf{s}^1, \mathbf{s}^2, \ldots, \mathbf{s}^{Ns}\}$, where Ns represents the number of test points over which the existence of an edge point will be verified, considering that the vector $\mathbf{s}^w$ contains the pixel coordinates $s_1^w = x_w$ and $s_2^w = y_w$.

The test **S** is generated by the Midpoint Circle Algorithm (MCA) [45]. The MCA aims to calculate the points $\mathbf{S} = \{\mathbf{s}^1, \mathbf{s}^2, \ldots, \mathbf{s}^{Ns}\}$ which are required to represent a circle considering the parameters (x_0, y_0) and r. See full details in [46].

Although the algorithm is considered to be the fastest providing a sub-pixel precision, it is important to assure it does not include points lying outside of the image plane. The matching function, also known as the objective function $J(\mathbf{C})$, represents the error resulting from pixels **S** of the circle candidate and **C** including only pixels that really exist in the edge image, yielding:

$$J(\mathbf{C}) = 1 - \frac{\sum_{v=1}^{Ns} V(\mathbf{s}^v)}{Ns} \qquad (6.6)$$

where $V(\mathbf{s}^v)$ is a function that verifies the pixel existence in $\mathbf{s}^v$, such as:

$$V(\mathbf{s}^v) = \begin{cases} 1 & \text{if the pixel in position } (x_v, y_v) \text{ exists} \\ 0 & \text{otherwise} \end{cases} \qquad (6.7)$$

Hence Eq. 6.6 accumulates the number of identified edge points (points in $\mathbf{S}$) that are actually present in the edge image. $\mathbf{Ns}$ is the number of pixels lying on the circle's perimeter that correspond to $\mathbf{C}$, currently under testing.

Therefore the algorithm tries to minimize $J(\mathbf{C})$, since a smaller value implies a better response (minimum error) of the "circularity" operator. The optimization process can thus be stopped either after the maximum number of epochs is reached or when the best individual is found.

6.2.3 EMO Implementation

The implementation of the presented algorithm based on EMO can be summarized into the following steps:

Step 1: The Canny filter is applied to find the edges, storing them into the $\mathbf{E} = \{\mathbf{e}^1, \mathbf{e}^2, \ldots, \mathbf{e}^{Np}\}$ vector. The *iteration* count is set to 0.

Step 2: m initial particles are generated (each one holding e_i, e_j and e_k elements, where e_i, e_j and $e_k \in \mathbf{E}$). Particles belonging to a seriously small or to a quite big radius are eliminated (collinear points are discarded). The objective function $J(\mathbf{C}^p)$ is evaluated to determine the best particle $\mathbf{C}^{best}$ (where $\mathbf{C}^{best} \leftarrow \text{argmin}\{J(\mathbf{C}^p), \forall p\}$).

Step 3: For a given coordinate $d \in (i, j, k)$, the particle $\mathbf{C}^p$ is assigned into a temporary point $\mathbf{t}$ to store the initial information. Next, a random number is selected and combined with δ yielding the step length. Therefore, the point $\mathbf{t}$ is moved along the direction d. The sign is determined randomly. If $J(\mathbf{t})$ is minimized, the particle $\mathbf{C}^p$ is replaced by $\mathbf{t}$ and the neighbourhood searching for a particle p finishes, otherwise $\mathbf{C}^p$ is held. Finally the *current best* particle $\mathbf{C}^{best}$ is updated.

Step 4: The charge between particles and its vector force are calculated according the Chap. 3. The particle with a better objective function holds a bigger charge and therefore a bigger attraction force.

Step 5: The new particle's position is calculated according Emo operators. $\mathbf{C}^{best}$ is not moved because it has the biggest force and it attracts others particles to itself.

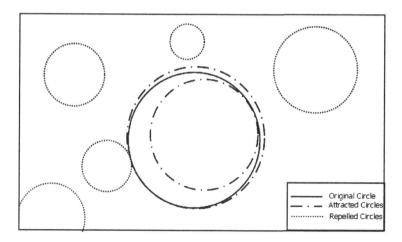

Fig. 6.2   An analogy to the Coulomb's law

Step 6:   The *Iteration* index is increased. If *Iteration* = *MAXITER* or if $J(\mathbf{C})$ value is as smaller as the pre-defined threshold value, then the algorithm is stopped and the flow jumps to step 7. Otherwise, it jumps to step 3.

Step 7:   The best $\mathbf{C}^{best}$ particle is selected from the last iteration.

Step 8:   From the original edge map, the algorithm marks points corresponding to $\mathbf{C}^{best}$. In case of multi-circle detection, it jumps to step 2.

Step 9:   Finally the best particle C_{Nc}^{best} from each circle is used to draw (over the original image) the detected circles, considering *Nc* as the number of circles actually found.

Figure 6.2 shows an analogy to the Coulomb's law. The original circle to be detected is represented by a solid line while the discontinuous line represents circles holding the most attractive force, i.e. they have the lowest error value. Other repelled circles are represented by dotted lines as they hold a larger error.

6.3   Experimental Results

In order to evaluate the performance of the circle detector presented in this chapter, several tests are implemented as follows: Sect. 6.3.1 Circle detection, Sect. 6.3.2 Shape discrimination, Sect. 6.3.3 Multiple circle detection, Sect. 6.3.4 Circular approximation, Sect. 6.3.5 Approximation from occluded or imperfect circles or arc detection and Sect. 6.3.6 Accuracy and computational time.

Each test is provided with a pool of particles $m = 10$, a maximum iteration for the local search *ITER* = 2, a step length for the local search $\delta = 3$. Parameters λ_1 and λ_2 are random values uniformly distributed while the maximum iteration value

MAXITER = 20. For the particles movement (Eq. 6.5), the step length λ is also set as a uniformly distributed random number. Finally, the search space (edge-only pixels) implies boundaries $u = 1$, $l = Np$ for each variable $\mathbf{e}_i$, $\mathbf{e}_j$ and $\mathbf{e}_k$.

6.3.1   Circle Localization

Synthetic images

The experimental setup includes the use of synthetic images of 256 × 256 pixels. Each image has been generated drawing only an imperfect circle (ellipse), which has been randomly located. Some of these images are contaminated adding noise to increase complexity in the detection process. The parameters to be detected are the centre of the circle position (x, y) and its radius (r). The algorithm is set to 20 iterations for each test image. For all the cases, the algorithm is able to detect the circle's parameters despite noise presence. The detection is robust to translation and scale, keeping a reasonable low elapsed time (typically under 1 ms). Figure 6.3 shows the results of the circle detection for two different synthetic images.

Natural images

One experiment exhibits the circle detection algorithm upon real-life images. Twenty five images of 640 × 480 pixels are used in the test. All images have been captured using an 8-bit colour digital camera and each natural scene includes a circle shape among other objects. All images are pre-processed using an edge detection algorithm before applying the EMO circle detector. Figure 6.4 shows two cases from 25 tested images.

(a)   **(b)**

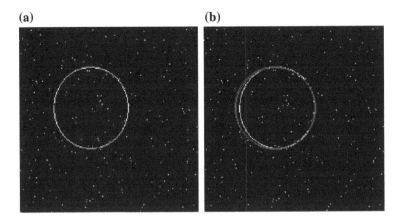

Fig. 6.3 Circle detection over synthetic images: **a** an original circle image with noise and **b** its corresponding detected circle

(a)   **(b)**

Fig. 6.4 Circle detection applied to two real-life images, **a** the detected *circle* is shown nearby the ring periphery; **b** the detected *circle* is shown nearby the ball's periphery

Real-life images rarely contain perfect circles. Therefore the detection algorithm approximates the circle that better adapts to the imperfect circle, i.e. the circle corresponding to the smallest error obtained by the objective function $J(\mathbf{C})$. All results have been statistically analyzed to ease comparison. The detection time for the image shown in Fig. 6.4a is 13.540807 s while it is 27.020633 s for the image in Fig. 6.4b. The detection algorithm has been executed over 20 times on the same image (Fig. 6.4b), yielding the same parameter set: $x_0 = 214$, $y_0 = 322$, and $r = 948$. The presented EMO algorithm is able to converge to the minimum solution as it is referred by the objective function $J(\mathbf{C})$. For this experiment 20 iterations have been required.

6.3.2 Test on Shape Discrimination

The section discusses on the ability for detecting circles whenever any other shape is present on the image. Five synthetic images of 540×300 pixels are considered for this experiment. Noise has been added to all images and a maximum of 20 iterations is used for the detection. Two examples of circle detection on synthetic images are shown by Fig. 6.5.

The same experiment is repeated over real-life images (see Fig. 6.6) confirming that the circle detection is completely feasible on natural real-life images despite other shapes being present on the scene.

(a)   **(b)**

(c)   **(d)**

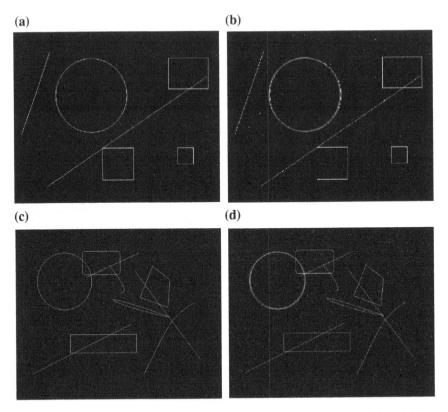

Fig. 6.5 Circle detection on synthetic images: images **a** and **c** are originals, while **b** and **d** show their detected *circles*

6.3.3 Multiple Circle Detection

The presented approach is also capable of detecting several circles over real-life images. First, a maximal number of circular shapes is set. The algorithm works on the original edge image until the first circle is detected. The first circle represents the circle with the minimum objective function value $J(\mathbf{C})$. This shape is then masked (eliminated) and the EMO circle detector operates over the modified image. This procedure is repeated until the maximum number of detected circular shapes is reached. Finally, a validation of all detected circles is performed by analyzing their circumference continuity as it is proposed in [27]. Such procedure becomes necessary in case more circular shapes might be required in the future, surpassing the number of circles currently detected in the image. In such a case, the system might provide a false statement: "no new circles are detected". On the other hand, the algorithm also seeks to identify any other circle-like shape in the image by selecting the best shapes until the maximum number of circular shapes is reached.

(a) **(b)**

(c) **(d)**

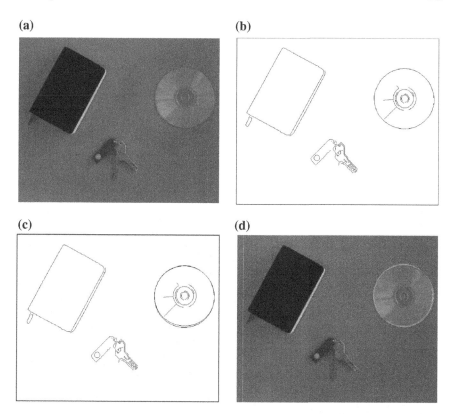

Fig. 6.6 Different shapes embedded into real-life images, **a** the test image, **b** the corresponding edge map, **c** the detected circle and **d** the detected circle over the original image

Figure 6.7a shows the edge image after applying the Canny algorithm and Fig. 6.7b presents the actual image including several detected circles which have been overlaid. The same has been done for other cases shown by Fig. 6.7c, d.

The EMO algorithm takes the optimized image from the previous step as input. In the last EMO iteration the output image, does not contain any complete circle, because it has been already detected and masked. The algorithm focuses on detecting other potential circles as a maximum of 20 iterations is commonly considered.

6.3.4 Circular Approximation

Since circle detection has been considered as an optimization problem, it is possible to approximate a shape as the concatenation of circles. This can be achieved by using the multiple-circle feature of the EMO algorithm (it has been explained in the

(a)   **(b)**

(c)   **(d)**

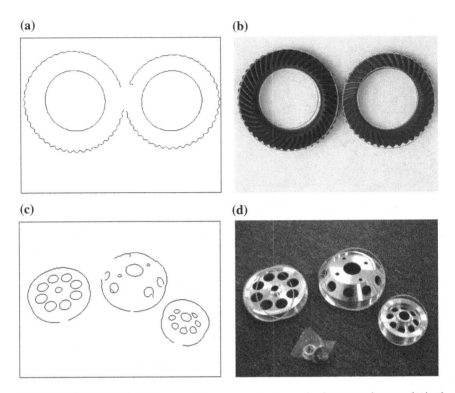

Fig. 6.7 Multiple circle detection on real images: **a** and **c** show edge images as they are obtained from applying the Canny algorithm. **b** and **d** present original images and the overlaid of detected circles is now evident

previous section). The EMO algorithm may continually find circles which may approach a given shape according to the values from the objective function $J(\mathbf{C})$.

Figure 6.8 shows some examples of circular approximation. Figure 6.8a shows a shape that has been built by the superposition of some circles. Figure 6.8b shows its circular approximation by 3 circles according to the EMO detector and finally Fig. 6.8c presents an ellipse that has been obtained by the concatenation of four circles shown by Fig. 6.8d.

6.3.5 Circle Localization from Occluded or Imperfect Circles and Arc Detection

Circle detection may also be useful to approximate circular shapes from arc segments, occluded circular segments or imperfect circles, which are common challenges for typical computer vision problems. The EMO algorithm may find the

(a)   **(b)**

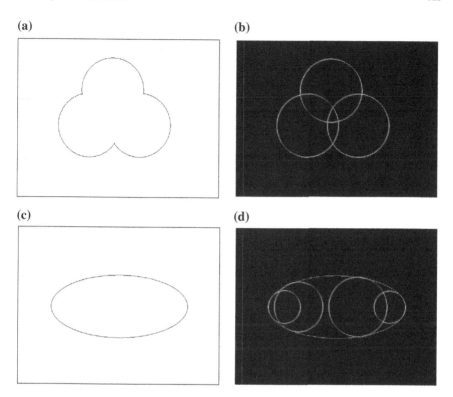

(c)   **(d)**

Fig. 6.8 Circular approximation: **a** the original image, **b** its circular approximation considering 3 circles, **c** original image, and **d** its circular approximation considering 4 circles

circle shape that approaches an arc according to the values in the objective function $J(\mathbf{C})$. Figure 6.9 shows some examples of this functionality.

6.3.6   Accuracy and Computational Time

This section aims to offer evidence on the algorithm's ability to find the best circle considering synthetic images with fixed noise. The experiment gets ten contaminated images of 256×256 pixels which contain only one circle centre at $x = 128$, $y = 128$, with $r = 64$ pixels. Two kinds of noise distributions are considered: the Salt and Pepper and the Gaussian noise.

The EMO algorithm iterates 35 times per image and the particle showing the best fitness value is regarded as the best circle matching the actual one in the image. The process is repeated over 25 times per image to test consistency. The accuracy evaluation is achieved by the Error sum (Es) that measures the difference between

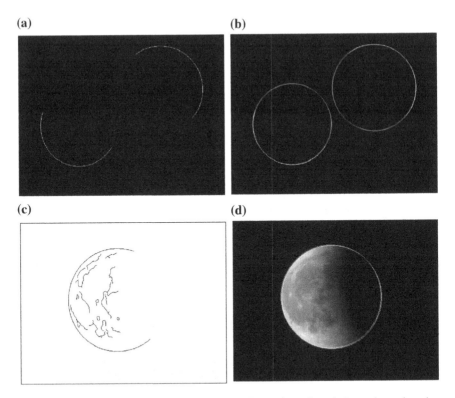

Fig. 6.9 Approximation of circles from occluded shapes, imperfect circles and arc detection: **a** original image with 2 arcs, **b** circle approximation for the first image, **c** occluded natural image of the moon and **d** circle approximation for the moon example

the ground truth circle (actual circle) and the detected one [7]. The error sum is defined as follows:

$$Es = |x_d - x_{true}| + |y_d - x_{true}| + |r_d - r_{true}|, \tag{6.8}$$

where $x_{true}, y_{true}, r_{true}$ are the coordinates of the centre and the radius value of the real circle in the image. Moreover x_d, y_d, r_d correspond to the values of centre and radius respectively once they have been detected by the method.

The first experiment considers contaminated images doped with Salt and Pepper noise. The EMO parameters are: maximum iterations $MAXITER = 35$; for the local search, $\delta = 4$ and $ITER = 4$. The added noise is produced by using MatLab©, considering noise levels between 1 and 10 %. The resulting values for **Es** and the computing elapsed time are reported in Table 6.1. Figure 6.10 shows three different images as examples, including 25 detected circles that are overlaid on each original image. It is evident for Fig. 6.10 that a higher added noise yields a higher dispersion on the detected shapes.

Table 6.1 Data analysis for images containing added Salt and Pepper noise

| Image properties | | | | Results | | | | | Computational time (s) | |
| | | | | Error sum (Es) | | | | | | |
Size	Circle center	Radius	Salt and pepper level	Total	Mean	Standard deviation	Mode		Mean	Standard deviation
256 × 256	(128, 128)	64	0.01	0	0	0	0		12.5505	0.5426
256 × 256	(128, 128)	64	0.02	0	0	0	0		13.0491	0.7455
256 × 256	(128, 128)	64	0.03	0	0	0	0		13.3401	0.7365
256 × 256	(128, 128)	64	0.04	0	0	0	0		14.3228	0.655
256 × 256	(128, 128)	64	0.05	2	0.0571	0.2355	0		14.2141	0.7982
256 × 256	(128, 128)	64	0.06	0	0	0	0		14.8669	0.9604
256 × 256	(128, 128)	64	0.07	3	0.0857	0.5071	0		15.3725	0.5147
256 × 256	(128, 128)	64	0.08	7	0.2	0.6774	0		14.5373	0.5147
256 × 256	(128, 128)	64	0.09	6	0.1714	0.5137	0		14.5747	0.8907
256 × 256	(128, 128)	64	0.10	28	0.8	2.1666	0		14.7288	0.8116

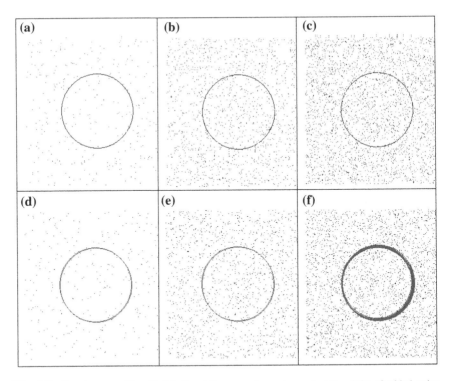

Fig. 6.10 Test images with added Salt and Pepper noise: **a** image holding 0.01 of added noise, **b** image including 0.05 of added noise, **c** image containing 0.1 of added noise, and **d**, **e** and **f** are the images showing the 25 *marked-circles* for each test image. All images resulted after applying the EMO algorithm

The second experiment explores the algorithm's application to images contaminated with Gaussian noise. By the definition, the Gaussian noise requires a threshold value to convert each value to binary pixels yielding an image exhibiting more noise than the Salt and Pepper contamination. The noise addition for the Gaussian noise falls between 1 and 10 %. The resulting values for *Es* and the computing elapsed time are reported in Table 6.2. Figure 6.11 shows three different images as examples, including the overlay of the 25 detected circles. It is evident again that dispersion on the detected circles increases proportionally to the added noise.

A similar test has been applied to different real images. However the Error sum (*Es*) is not used because the coordinates of the centre and the radio for each circle are unknown. Table 6.3 shows the results after applying the EMO circle detector considering the natural image algorithm presented in Fig. 6.12. The resulting 25 detected-circles are also overlaid on the original images.

Table 6.2 Data analysis following the Gaussian noise application

Image properties					Results						
					Error sum (Es)				Computational time (s)		
Size	Circle center	Radius	Gaussian noise		Total	Mean	Standard deviation	Mode	Mean	Standard deviation	
			Mean	Standard deviation							
256 × 256	(128, 128)	64	0	0.01	0	0	0	0	11.7008	0.6566	
256 × 256	(128, 128)	64	0	0.02	0	0	0	0	12.8182	0.9257	
256 × 256	(128, 128)	64	0	0.03	0	0	0	0	13.8073	0.6438	
256 × 256	(128, 128)	64	0	0.04	6	0.1714	0.5681	0	13.7866	0.8309	
256 × 256	(128, 128)	64	0	0.05	10	0.2857	0.825	0	14.0476	1.4691	
256 × 256	(128, 128)	64	0	0.06	26	0.7429	0.95	0	14.0622	0.5463	
256 × 256	(128, 128)	64	0	0.07	32	0.9143	1.961	0	14.6079	0.4346	
256 × 256	(128, 128)	64	0	0.08	158	4.5143	11.688	0	14.1732	0.8198	
256 × 256	(128, 128)	64	0	0.09	106	3.0286	6.5462	0	13.9074	0.8165	
256 × 256	(128, 128)	64	0	0.1	175	5	8.7447	0	15.1276	0.8606	

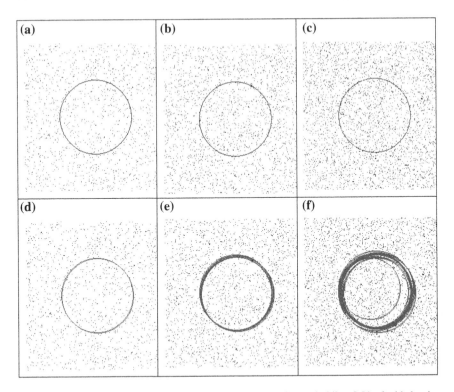

Fig. 6.11 Test images contaminated with Gaussian noise: **a** image holding 0.01 of added noise, **b** image including 0.05 of added noise, **c** image containing 0.1 of added noise while images **d**, **e** and **f** are showing the 25 *marked-circles* for each test image. All images resulted after applying the EMO algorithm

Table 6.3 Experimental results after applying the EMO circle detector to real-life images

Image properties		Results					
		Matching error (%)				Computational time (s)	
Image name	Size	Total	Mean	Standard deviation	Mode	Mean	Standard deviation
Cue ball	430 × 473	28.1865	0.805	0.0061	0.8035	26.6641	1.4018
Street lamp	474 × 442	19.0638	0.545	0.0408	0.5326	23.4763	1.7857
Wheel	640 × 480	22.265	0.636	0.0062	0.6336	31.356	1.4351

(a)   (b)

(c)

Fig. 6.12 Real-life images used in the experiments, **a** Cue ball, **b** Wheel, **c** Street ball, including the overlaid of detected circles

6.4   Conclusions

This chapter presents an algorithm for the automatic detection of circular shapes among cluttered and noisy images with no consideration of the conventional Hough transform principles. This work describes a circle detection method which is based on a nature-inspired technique known as the Electromagnetism-Like Optimization (EMO). It is a heuristic method for solving complex optimization problems that has been inspired by electromagnetism principles. To the best of our knowledge, the EMO has not been applied to any image processing task until date. The algorithm uses the encoding of three non-collinear edge points as candidate circles (x, y, r) in the edge-only image of the scene. An objective function evaluates if these candidate circles (charged particles) are really present. Guided by the values of this objective function, the set of encoded candidate circles are evolved using the EMO algorithm so that they can fit into the actual circles on the edge-only map of the image. As it can be perceived from the result, our approach detects the circle in complex images with little visual distortion despite the presence of noisy background pixels and still showing good accuracy and consistency.

An important feature of this work is to consider the circle detection problem as an optimization approach. Such view enables the EMO algorithm to find circle parameters according to $J(\mathbf{C})$ instead of voting all possible chances for occluded or imperfect circles as other methods do.

Although Hough Transform methods for circle detection also use three edge points to cast one vote for the potential circular shape in the parameter space, they would require huge amounts of memory and longer computational times to obtain a sub-pixel resolution. In the HT-based methods, the parameter space is quantized and the exact parameters for a valid circle are often not equal to the quantized parameters. This evidently yields a non-exact determination for a circle actually present in the image. However, the presented EMO method does not employ the quantization of the parameter space. In our approach, the detected circles are directly obtained from Eqs. 6.2 to 6.5 effectively detecting the circle with a sub-pixel accuracy.

Although the results offer evidence to demonstrate that the EMO method can yield good results on complicated and noisy images, the aim of our chapter is not to devise a circle-detection algorithm that could beat all circle detectors currently available, but to show that Electromagnetism-Like systems can effectively be considered as an attractive alternative for detecting circular shapes.

References

1. Liu, J., Tsui, K.: Toward nature-inspired computing. Commun. ACM **49**(10), 59–64 (2006)
2. Hongwei, M.: Handbook of Research on Artificial Immune Systems and Natural Computing: Applying Complex Adaptive Technologies. IGI Global, United States of America (2009)
3. Lévy, P.: From social computing to reflexive collective intelligence: the IEML research program. Inf. Sci. **180**(1), 71–94 (2010)
4. Gruber, T.: Collective knowledge systems: where the social web meets the semantic web. Web Semant: Sci, Serv Agents World Wide Web **6**(1), 4–13 (2008)
5. Teodorović, D.: Swarm intelligence systems for transportation engineering: principles and applications. Transp. Res. Part C: Emerg. Technol. **16**(6), 651–667 (2008)
6. Karaboga, D., Akay, B.: A comparative study of Artificial Bee Colony algorithm. Appl. Math. Comput. **214**(1), 108–132 (2009)
7. Blum, C.: Ant colony optimization: introduction and recent trends. Phys. Life Rev. **2**(4), 353–373 (2005)
8. Naji-Azimi, Z., Toth, P., Galli, L.: An electromagnetism metaheuristic for the unicost set covering problem. Eur. J. Oper. Res. **205**(2), 290–300 (2010)
9. da Fontoura Costa, L., Marcondes Cesar, R., Jr.: Shape Análisis and Classification. CRC Press, Boca Raton FL. (2001)
10. Davies, E.R.: Machine Vision: Theory, Algorithms, Practicalities. Academic Press, London (1990)
11. Yuen, H., Princen, J., Illingworth, J., Kittler, J.: Comparative study of Hough transform methods for circle finding. Image Vision Comput. **8**(1), 71–77 (1990)

12. Iivarinen, J., Peura, M., Sarela, J., Visa, A.: Comparison of combined shape descriptors for irregular objects. In: Proceedings of 8th British Machine Vision Conference, pp. 430–439. Cochester, UK (1997)
13. Jones, G., Princen, J., Illingworth, J., Kittler, J. Robust estimation of shape parameters. In: Proc. British Machine Vision Conf., pp. 43– 48. (1990)
14. Fischer, M., Bolles, R.: Random sample consensus: a paradigm to model fitting with applications to image analysis and automated cartography. CACM **24**(6), 381–395 (1981)
15. Bongiovanni, G., Crescenzi, P.: Parallel simulated annealing for shape detection. Comput. Vis. Image Underst. **61**(1), 60–69 (1995)
16. Roth, G., Levine, M.D.: Geometric primitive extraction using a genetic algorithm. IEEE Trans. Pattern Anal. Mach. Intell. **16**(9), 901–905 (1994)
17. Peura, M., Iivarinen, J.: Efficiency of simple shape descriptors. In: Arcelli, C., Cordella, L.P., di Baja, G.S. (eds.) Advances in Visual Form Analysis, pp. 443–451. World Scientific, Singapore (1997)
18. Muammar, H., Nixon, M.: Approaches to extending the Hough transform. In: Proceedings of International Conference on Acoustics, Speech and Signal Processing ICASSP_89, vol. 3, pp. 1556–1559 (1989)
19. Atherton, T.J., Kerbyson, D.J.: Using phase to represent radius in the coherent circle Hough transform. In: Proceedings of IEEE Colloquium on the Hough Transform, IEEE, London (1993)
20. Shaked, D., Yaron, O., Kiryati, N.: Deriving stopping rules for the probabilistic Hough transform by sequential analysis. Comput. Vision Image Underst. **63**, 512–526 (1996)
21. Xu, L., Oja, E., Kultanen, P.: A new curve detection method: randomized Hough transform (RHT). Pattern Recogn. Lett. **11**(5), 331–338 (1990)
22. Han, J.H., Koczy, L.T., Poston, T.: Fuzzy Hough transform. In: Proceedings of 2nd International Conference on Fuzzy Systems, vol. 2, pp. 803–808 (1993)
23. Becker, J., Grousson, S., Coltuc, D.: From Hough transforms to integral transforms. In: Proceedings of International Geoscience and Remote Sensing Symposium, 2002 IGARSS_02, vol. 3, pp. 1444–1446 (2002)
24. Lutton, E., Martinez, P.: A genetic algorithm for the detection 2-D geometric primitives on images. In: Proceedings of the 12th International Conference on Pattern Recognition, vol. 1, pp. 526–528 (1994)
25. Yao, J., Kharma, N., Grogono, P.: Fast robust GA-based ellipse detection. In: Proceedings of 17th International Conference on Pattern Recognition ICPR-04, vol. 2, pp. 859–862. Cambridge, UK (2004)
26. Yuen, S., Ma, C.: Genetic algorithm with competitive image labelling and least square. Pattern Recogn. **33**, 1949–1966 (2000)
27. Ayala-Ramirez, V., Garcia-Capulin, C.H., Perez-Garcia, A., Sanchez-Yanez, R.E.: Circle detection on images using genetic algorithms. Pattern Recogn. Lett. **27**, 652–657 (2006)
28. Rosin, P.L., Nyongesa, H.O.: Combining evolutionary, connectionist, and fuzzy classification algorithms for shape analysis. In: Cagnoni, S. et al. (eds.) Proceedings of EvoIASP, Real-World Applications of Evolutionary Computing, pp. 87–96 (2000)
29. Rosin, P.L.: Further five point fit ellipse fitting. In: Proceedings of 8th British Machine Vision Conference, pp. 290–299. Cochester, UK (1997)
30. Zhang, X., Rosin, P.L.: Superellipse fitting to partial data. Pattern Recogn. **36**, 743–752 (2003)
31. Andrei, N.: Acceleration of conjugate gradient algorithms for unconstrained optimization. Appl. Math. Comput. **213**(2), 361–369 (2009)
32. Zhang, Q., Mahfouf, M.: A nature-inspired multi-objective optimization strategy based on a new reduced space search ing algorithm for the design of alloy steels. Eng. Appl. Artif. Intell. (2010). doi:10.1016/j.engappai.2010.01.017
33. Birbil, SÍ., Fang, S.-C.: An electromagnetism-like mechanism for global optimization. J. Global Optim. **25**, 263–282 (2003)

34. Gaafar, L.K., Masoud, S.A., Nassef, A.O.: A particle swarm-based genetic algorithm for scheduling in an agile environment. Comput. Ind. Eng. **55**(3), 707–720 (2008)
35. Chen, Y.-P., Jiang, P.: Analysis of particle interaction in particle swarm optimization. Theoret. Comput. Sci. **411**(21), 2101–2115 (2010)
36. Maniezzo, V., Carbonaro, A.: Ant Colony Optimization: An Overview. Essays and Surveys in Metaheuristics, pp. 21–44. Kluwer Academic Publisher, Dordrecht (1999)
37. Wu, P., Yang, W.-H., Wei, N.-C.: An electromagnetism algorithm of neural network analysis— an application to textile retail operation. J. Chin. Inst. Ind. Eng. **21**(1), 59–67 (2004)
38. Tsou, C.-S., Kao, C.-H.: Multi-objective inventory control using electromagnetism-like metaheuristic. Int. J. Prod. Res. **46**(14), 3859–3874 (2008)
39. Rocha, A., Fernandes, E.: Hybridizing the electromagnetism-like algorithm with descent search for solving engineering design problems. Int. J. Comput. Math. **86**(10), 1932–1946 (2009)
40. Rocha, A., Fernandes, E.: Modified movement force vector in an electromagnetism-like mechanism for global optimization. Optim. Meth. Softw. **24**(2), 253–270 (2009)
41. Birbil, Sİ., Fang, S.-C., Sheu, R.L.: On the convergence of a population-based global optimization algorithm. J. Global Optim. **30**(2), 301–318 (2004)
42. Naderi, B., Tavakkoli-Moghaddam, R., Khalili, M.: Electromagnetism-like mechanism and simulated annealing algorithms for flowshop scheduling problems minimizing the total weighted tardiness and makespan. Knowl.-Based Syst. **23**(2), 77–85 (2010)
43. Yurtkuran, A., Emel, E.: A new hybrid electromagnetism-like algorithm for capacitated vehicle routing problems. Expert Syst. Appl. **37**(4), 3427–3433 (2010)
44. Jhang, J.-Y., Lee, K.-C.: Array pattern optimization using electromagnetism-like algorithm. AEU Int. J. Electron. Commun. **63**(6), 491–496 (2009)
45. Bresenham, J.E.: A linear algorithm for incremental digital display of circular arcs. Commun. ACM **20**, 100–106 (1977)
46. Van Aken, J.R.: An efficient ellipse drawing algorithm. CG&A, **4**(9), 24–35 (1984)

Chapter 7
A Medical Application: Blood Cell Segmentation by Circle Detection

7.1 Introduction

Evolutionary computing is a field of research that is concerned with both the use of biology as an inspiration for solving computational problems and the use of the natural physical phenomena to solve real world problems. Moreover, nature-inspired computing has proved to be useful in various application areas [1] with relevant contributions in optimization, pattern recognition, shape detection and machine learning. In particular they have recently gained considerable research interest from the computer vision community as it they have successfully contributed to solve challenging computer vision problems.

On the other hand, White Blood Cells (WBC) also known as leukocytes play a significant role in the diagnosis of different diseases. Although digital image processing techniques have successfully contributed to analyze the cells leading to more accurate and reliable systems for disease diagnosis, a high variability on cell shape, size, edge and localization complicates the data extraction process. Moreover, the contrast between cell boundaries and the image's background varies according to lighting conditions during the capturing process.

Many works have been conducted in the area of blood cell detection, for instance [2–5]. One of the latest advances in white blood cell detection research is the algorithm proposed by Wang [6] which is based on the fuzzy cellular neural network (FCNN). Although such method has proved successful in detecting only one leukocyte in the image, it has not been tested over images containing several white cells. Moreover, its performance commonly decays when the iteration number is not properly defined which represents a challenging problem itself with no clear clues on how to make the best choice.

Since blood cells can be approximated with a quasi-circular form, a circular detector algorithm may be applied. The problem of detecting circular features holds paramount importance for image analysis, in particular for medical image analysis [7]. The circle detection in digital images is commonly performed by the Circular

© Springer International Publishing AG 2017
D. Oliva and E. Cuevas, *Advances and Applications of Optimised Algorithms in Image Processing*, Intelligent Systems Reference Library 117,
DOI 10.1007/978-3-319-48550-8_7

Hough Transform [8]. A typical Hough-based approach employs an edge detector whose information guides the inference for circle locations and radius values. Peak detection is then performed by averaging, filtering and histogramming the transform space. However, such approach requires a large storage space given the required 3-D cells to cover all parameters (x, y, r). It also implies a high computational complexity yielding a low processing speed. The accuracy of the extracted parameters for the detected circle is poor, particularly in presence of noise [9]. For a digital image holding a significant width and height and a densely populated edge pixel map, the required processing time for Circular Hough Transform makes it prohibitive to be deployed in real time applications. In order to overcome such a problem, some other researchers have proposed new approaches based on the Hough transform, for instance the probabilistic Hough transform [10, 11], the randomized Hough transform (RHT) [12] and the fuzzy Hough transform [13]. Alternative transformations have also been presented in the literature as the one proposed by Becker in [14]. Although those new approaches demonstrated better processing speeds, in comparison to the original Hough Transform, they are still very sensitive to noise.

As an alternative to Hough Transform-based techniques, the problem of shape recognition in computer vision has also been handled through optimization methods. In particular, Genetic Algorithms (GA) has recently been used for important shape detection tasks e.g. Roth and Levine proposed use of GA for primitive extraction of images [15]. Lutton et al. carried out a further improvement of the aforementioned method [16]. Yao et al. came up with a multi-population GA to detect ellipses [17]. In [18], GA was used for template matching when the pattern has been the subject of an unknown affine transformation. Ayala–Ramirez et al. presented a GA based circle detector [19] which is capable of detecting multiple circles on real images but fails frequently to detect imperfect circles. Another example is presented in [20] where it employs the electromagnetism-like algorithm (EMO) as optimization procedure. In the case of ellipsoidal detection, Rosin proposes in [21] an ellipse fitting algorithm that uses five points. In Zhang and Rosin [22] extends the later algorithm to fit data into super-ellipses. Most of those approaches allow the circle detection offering an acceptable computation time even under noisy conditions. However, they still have failed at complex conditions such as occlusion or superposition.

Although detection algorithms based on optimization approaches present several advantages in comparison to those based on the Hough Transform, they have been scarcely applied to WBC detection. One exception is the work presented by Karkavitsas and Rangoussi [23] that solves the WBC detection problem through the use of GA. However, since the evaluation function, which assesses the quality of each solution, considers the number of pixels contained inside of a circle with fixed radius, the method is prone to produce misdetections particularly for images that contained overlapped or irregular WBC.

In this chapter, the blood cell detection task is considered as an optimization problem and the EMO algorithm is used to build the circular approximation. The EMO algorithm [24] is a stochastic evolutionary computation technique based on the electromagnetism theory. It considers each solution to be a charged particle. The charge of each particle is determined by an objective function. Thereby, EMO moves each particle according to its charge within an attraction or repulsion field among the population using the Coulomb's law and the superposition principle. This attraction–repulsion mechanism of the EMO algorithm corresponds to the reproduction, crossover and mutation in GA [25]. In general, the EMO algorithm can be considered as a fast and robust algorithm representing an alternative to solve complex, nonlinear, non-differentiable and non-convex optimization problems. The principal advantages of the EMO algorithm lies on several facts: it has no gradient operation, it can be used directly on a decimal system, it needs only few particles to converge and the convergence existence has been already verified [26].

The EMO-based circle detector uses the encoding of three edge points that represent candidate circles in the edge map of the scene. The quality of each individual is calculated by using an objective function which evaluates if such candidate circles are really present in the edge map of the image. The better a candidate circle approximates the actual edge-circle, the objective function value decreases. Therefore, the detection performance depends on the quality of the edge map as it is obtained from the original images. However, since smear images present different imaging conditions and staining intensities, they produce edge maps partially damaged by noisy pixels. Under such conditions, the use of the EMO-based circle detector can not be directly applied to WBC detection.

This chapter presents an algorithm for the automatic detection of blood cell images based in the EMO algorithm. The proposed method uses the encoding of three non-collinear edge points as candidate circles (x, y, r) in the edge-only image of the scene. Guided by the values of the objective function, the set of encoded candidate circles are evolved using the EMO algorithm so that they can fit into actual circles on the edge map of the image. The approach generates a sub-pixel shape detector which can effectively identify leukocytes in real images despite the WBC exhibiting a significant occluded portion. Experimental evidence shows the effectiveness of such method in detecting leukocytes under different conditions. Comparison to the state-of-the-art Wang's algorithm [6] on multiple images, demonstrates a better performance of the proposed method.

The main contribution of this study is the proposal of a new WBC detector algorithm that efficiently recognize WBC under different complex conditions while considering the whole process as an circle detection problem. Although circle detectors based on optimization present several interesting properties, to the best of our knowledge, they have not yet been applied to any medical image processing up to date.

This chapter is organized as follows: in Sect. 7.2 the EMO detection approach is formulated. Section 7.3 presents a numerical example, Sect. 7.4 presents the experimental results and Sect. 7.5 compares the EMO detector to other relevant methods. Finally, Sect. 7.6 discusses on some conclusions and future work.

7.2   Circle Detection

This section introduces the basic concepts of circle detection using EMO to find the best shape on a digital image.

7.2.1   Data Preprocessing

In order to detect circle shapes, candidate images must be preprocessed first by the well-known Canny algorithm which yields a single-pixel edge-only image. Then, the (x_i, y_i) coordinates for each edge pixel p_i are stored inside the edge vector $P = \{p_1, p_2, \ldots, p_{N_p}\}$, with N_p being the total number of edge pixels.

7.2.2   Particle Representation

In order to construct each particle C_a (circle candidate), the indexes e_1, e_2 and e_3, which represent three edge points previously stored inside the vector P, must be grouped assuming that the circle's contour connects them. Therefore, the circle $C_a = \{p_{e_1}, p_{e_2}, p_{e_3}\}$ passing over such points may be considered as a potential solution for the detection problem. Considering the configuration of the edge points shown by Fig. 7.1, the circle center (x_0, y_0) and the radius r of C_a can be characterized as follows:

$$(x - x_0)^2 + (y - y_0)^2 = r^2 \tag{7.1}$$

Fig. 7.1 Circle candidate (charged-particle) built from the combination of points p_{e_1}, p_{e_2} and p_{e_3}

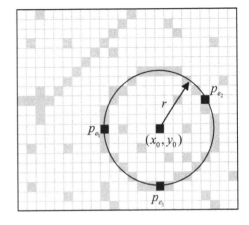

where x_0 and y_0 are computed through the following equations:

$$x_0 = \frac{\det(\mathbf{A})}{4((x_{e_2} - x_{e_1})(y_{e_3} - y_{e_1}) - (x_{e_3} - x_{e_1})(y_{e_2} - y_{e_1}))},$$
$$y_0 = \frac{\det(\mathbf{B})}{4((x_{e_2} - x_{e_1})(y_{e_3} - y_{e_1}) - (x_{e_3} - x_{e_1})(y_{e_2} - y_{e_1}))},$$

(7.2)

with $\det(\mathbf{A})$ and $\det(\mathbf{B})$ representing determinants of matrices $\mathbf{A}$ and $\mathbf{B}$ respectively, considering:

$$\mathbf{A} = \begin{bmatrix} x_{e_2}^2 + y_{e_2}^2 - (x_{e_1}^2 + y_{e_1}^2) & 2 \cdot (y_{e_1} - y_{e_1}) \\ x_{e_3}^2 + y_{e_3}^2 - (x_{e_1}^2 + y_{e_1}^2) & 2 \cdot (y_{e_3} - y_{e_1}) \end{bmatrix}$$
$$\mathbf{B} = \begin{bmatrix} 2 \cdot (x_{e_2} - x_{e_1}) & x_{e_2}^2 + y_{e_2}^2 - (x_{e_1}^2 + y_{e_1}^2) \\ 2 \cdot (x_{e_3} - x_{e_1}) & x_{e_3}^2 + y_{e_3}^2 - (x_{e_1}^2 + y_{e_1}^2) \end{bmatrix},$$

(7.3)

the radius r can therefore be calculated using:

$$r = \sqrt{(x_0 - x_{e_d})^2 + (y_0 - y_{e_d})^2}$$

(7.4)

where $d \in \{1, 2, 3\}$, and (x_{e_d}, y_{e_d}) are the coordinates of any of the three selected points which define the action C_a. Figure 7.1 illustrates main parameters defined by Eqs. (7.1–7.4).

The shaping parameters for the circle, $[x_0, y_0, r]$ can be represented as a transformation T of the edge vector indexes e_1, e_2 and e_3.

$$[x_0, y_0, r] = T(e_1, e_2, e_3)$$

(7.5)

By exploring each index as a particle, it is possible to sweep the continuous space searching for shape parameters by means of the EMO algorithm. This approach reduces the search space by eliminating unfeasible solutions.

7.2.3 Objective Function

In order to model the fitness function, the circumference coordinates of the circle candidate C_a are calculated as a virtual shape which must be validated, i.e. if it really exists in the edge image. The circumference coordinates are grouped within the test set $H_a = \{h_1, h_2, \ldots, h_{N_s}\}$, with N_s representing the number of points over which the existence of an edge point, corresponding to C_a, should be verified.

The test H_a is generated by the midpoint circle algorithm (MCA) [27] which is a well-known algorithm to determine the required points for drawing a circle. MCA requires as inputs only the radius r and the center point (x_0, y_0) considering only the first octant over the circle equation $x^2 + y^2 = r^2$. It draws a curve starting at point $(r, 0)$ and proceeds upwards-left by using integer additions and subtractions. The MCA aims to calculate the required points H_a in order to represent a circle candidate. Although the algorithm is considered as the quickest providing a sub-pixel precision, it is important to assure that points lying outside the image plane must not be considered in H_a.

The objective function $J(C)$ represents the matching error produced between the pixels H_a of the circle candidate C_a (particle) and the pixels that actually exist in the edge-only image, yielding:

$$J(C) = 1 - \frac{\sum_{v=1}^{Ns} E(h_v)}{N_s} \qquad (7.6)$$

where $E(h_v)$ is a function that verifies the pixel existence in h_v, being $h_v \in H_a$ and N_s is the number of pixels lying over the perimeter and corresponding to C_a, currently under testing. Hence the function $E(h_v)$ is defined as:

$$E(h_v) = \begin{cases} 1 & \text{if the test pixel } h_v \text{ is an edge point} \\ 0 & \text{otherwise} \end{cases} \qquad (7.7)$$

A value of $J(C)$ near to zero implies a better response from the "circularity" operator. Figure 7.2 shows the procedure to evaluate a candidate action C_a with its representation as a virtual shape H_a. Figure 7.2a shows the original edge map, while Fig. 7.2b presents the virtual shape H_a representing the action $C_a = \{p_{e_1}, p_{e_2}, p_{e_3}\}$. In Fig. 7.2c, the virtual shape H_a is compared to the original image, point by point, in order to find coincidences between virtual and edge points. The action has been built from points p_{e_1}, p_{e_2} and p_{e_3} which are shown by Fig. 7.2a. The virtual shape H_a, obtained by MCA, gathers 56 points ($N_s = 56$) with only 18 of them existing in both images (shown as blue points plus red points in Fig. 7.2c) and yielding: $\sum_{h=1}^{Ns} E(h_v) = 18$, therefore $J(C_a) \approx 0.67$.

7.2.4   EMO Implementation

The implementation of the proposed algorithm can be summarized into the following steps:

Step 1 The Canny filter is applied to find the edges and store them in the $P = \{p_1, p_2, \ldots, p_{N_p}\}$ vector. The index k is set to 1.

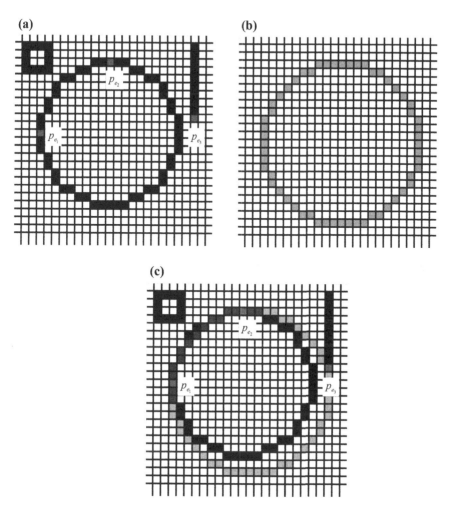

Fig. 7.2 Procedure to evaluate the objective function $J(C_a)$: The image shown by (**a**) presents the original edge image while (**b**) portraits the virtual shape H_a corresponding to C_a. The image in (**c**) shows coincidences between both images through *blue* or *red pixels* while the virtual shape is also depicted in *green*

Step 2: m initial particles are generated ($C_{a,1}, a \in [1, m]$). Particles belonging to a seriously small or to a quite big radius are eliminated (collinear points are discarded).

Step 3: The objective function $J(C_{a,k})$ is evaluated to determine the best particle C^B (where $C^B \leftarrow \arg\min\{J(C_{a,k})\}$).

Step 4: The charge between particles and its vector force is calculated according the method presented in Chap. 3. The particle with a better objective function holds a bigger charge and therefore a bigger attraction force.

Step 5: The particles are moved according to their force magnitude. The new particle's position C_a^y is calculated according with Chap. 3. C^B is not moved because it has the biggest force and it attracts others particles to itself.

Step 6: For each C_a^y a maximum of I_l points are generated in each coordinate direction in the δ neighborhood of C_a^y. This means that the process of generating local points is continued for each C_a^y until either a better C_a^z is found or the $n \times I_l$ trial is reached.

Step 7: The new particles $C_{a,k+1}$ are selected from C_a^y and C_a^z.

Step 8: The k index is increased. If $k = MAXITER$ or if $J(C_{a,k})$ value is as smaller as the pre-defined threshold value then the algorithm is stopped and the flow jumps to step 9. Otherwise, it jumps to step 3.

Step 9: The best C^B particle is selected from the last iteration.

Step 10: From the original edge map, the algorithm marks points corresponding to C^B. In case of multi-circle detection, it jumps to step 2.

Step 11: Finally the best particle C_{Nc}^B from each circle is used to draw (over the original image) the detected circles, considering Nc as the number of circles actually found.

Figure 7.3 shows an analogy to the Coulomb's law. The original figures to be detected are represented by a solid black line while the shapes with discontinuous gray lines represent the candidate circles. Since the candidate circles $C_{1,k}$ and $C_{3,k}$ present a high value in the fitness function $J\left(C_{a,k}\right)$, they are repelled (blue lines), moving away the shapes. In contrast the circle candidate $C_{2,k}$ that holds a small value of $J\left(C_{a,k}\right)$, is attracted (red line) to the circular shape contained in the image.

7.2.5   White Blood Cell Detection

This section describes the process developed to detect white blood cells combining the EMO-based circle detector previously presented with a new objective function.

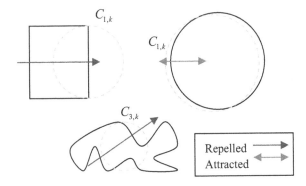

Fig. 7.3 An analogy to the Coulomb's law

Image preprocessing

To employ the proposed detector, smear images must be preprocessed to obtain two new images: the segmented image and its corresponding edge map. The segmented image is produced by using a segmentation strategy whereas the edge map is generated by a border extractor algorithm. Both images are considered by the new objective function to measure the resemblance of a candidate circle with an actual WBC.

The goal of the segmentation strategy is to isolate the white blood cells (WBC's) from other structures such as red blood cells and background pixels. Information of color, brightness and gradients are commonly used within a thresholding scheme to generate the labels to classify each pixel. Although a simple histogram thresholding can be used to segment the WBC's, at this work the Diffused Expectation-Maximization (DEM) has been used to assure better results [28].

DEM is an Expectation-Maximization (EM) based algorithm which has been used to segment complex medical images [29]. In contrast to classical EM algorithms, DEM considers the spatial correlations among pixels as a part of the minimization criteria. Such adaptation allows to segment objects in spite of noisy and complex conditions.

For the WBC's segmentation, the DEM has been configures considering three different classes ($K = 3$), $g(\nabla h_{ik}) = |\nabla h_{ik}|^{-9/5}$, $\lambda = 0.1$ and $m = 10$. These values have been found as the best configuration set according to [28]. As a final result of the DEM operation, three different thresholding points are obtained: the first corresponds to the WBC's, the second to the red blood cells whereas the third represents the pixels classified as background. Figure 7.4b presents the segmentation results obtained by the DEM approach employed at this work considering the Fig. 7.4a as the original image.

Once the segmented image has been produced, the edge map is computed. The purpose of the edge map is to obtain a simple image representation that preserves object structures. Optimization-based circle detectors [23–26] operate directly over the edge map in order to recognize circular shapes. Several algorithms can be used to extract the edge map; however, at this work, the morphological edge detection procedure [30] has been used to accomplish such a task. Morphological edge detection is a traditional method to extract borders from binary images in which original images (I_B) are eroded by a simple structure element (I_E). Then, the eroded image is inverted $\overline{(I_E)}$ and compared with the original image $(\overline{I_E} \wedge I_B)$ in order to detect pixels which are present in both images. Such pixels compose the computed edge map from I_B. Figure 7.4c shows the edge map obtained by using the morphological edge detection.

The modified EMO based circle detector

The circle detection approach uses the encoding of three edge points that represent candidate circles in the image. In the original EMO-based circle detector, the quality of each individual is calculated by using an objective function which evaluates the existence of a candidate circle considering only information from the edge map (shape structures). The better a candidate circle approximates the actual

(a)   **(b)**

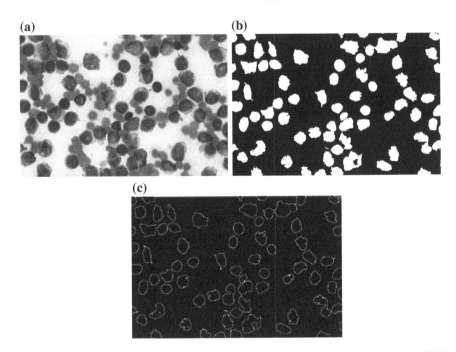

(c)

Fig. 7.4 Preprocessing process. **a** original smear image, **b** segmented image obtained by DEM and **c** the edge map obtained by using the morphological edge detection procedure

edge-circle, the objective function value decreases. Therefore, the detection performance depends on the quality of the edge map that is obtained from the original images. However, since smear images present different imaging conditions and staining intensities, they produce edge maps partially damaged by noisy pixels. Under such conditions, the use of the EMO-based circle detector can not be directly applied to WBC detection.

In order to use the EMO-based circle detector within the context of WBC detection, it is necessary to change the fitness function presented in Eq. 7.6. At this work, a new objective function has been derived to measure the resemblance of a candidate circle to an actual WBC based on the information from the edge map and the segmented image. Such new objective function takes into consideration not only the information provided by the edge map, but also the relationship among the pixels falling inside of the candidate circle which is contained in the segmented image, validating the existence of the WBC. This new function $J(\mathbf{C})$ is thus calculated as follows:

$$J_{New}(\mathbf{C}) = 2 - \frac{\sum\limits_{v=1}^{N_s} E(h_v)}{N_s} - \frac{Wp}{Bp} \qquad (7.8)$$

where h_v and N_s keep the same meaning than Eq. 7.6. *Wp* is the amount of white pixels falling inside the candidate circle represented by **C**. Likewise, *Bp* corresponds to the total number of black pixels falling inside **C**.

To illustrate the functionality of the new objective function, Fig. 7.5 presents a detection procedure which considers a complex image. Figure 7.5a shows the original smear image containing a WBC and a stain produced by the coloring process. Figures 7.5b and c represent the segmented image and the edge map, respectively. Since the stain contained in the smear image (Fig. 7.5a) possesses similar properties than a WBC, it remains as a part of the segmented image (Fig. 7.5b) and the edge map (Fig. 7.5c). Such an inconsistency produces big detection errors in case the EMO-based circle detector is used without modification. Figure 7.5d presents detection results obtained by the original EMO-based circle detector. As the original objective function considers only the number of coincidences between the candidate circle and the edge map, circle candidates that match with a higher number of edge pixels are chosen as the best circle instances. In Fig. 7.5d, the detected circle presents a coincidence of 37 different pixels in the edge map. Such coincidence is considered as the best possible under the restrictions of the original objective function. On the other hand, when the modified objective function is used in the recognition procedure, the accuracy and the robustness of the detection are both significantly improved. By using the new objective function, information from the segmented image is employed to refine the solution that is provided by coincidences with the edge map. Figure 7.5e presents the detection result that has been produced by the modified EMO-based circle detector. In the

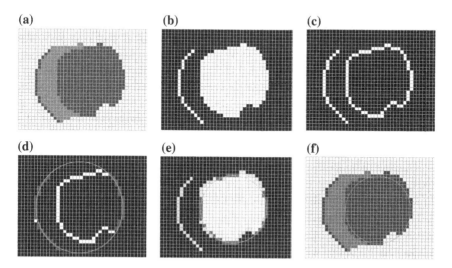

Fig. 7.5 WBC detection procedure. **a** Smear image. **b** Segmented image. **c** Edge map. **d** Detected circle by using the original objective function. *Red points* show the coincidences between the candidate circle and the edge map. **e** Detected circle by using the new objective function. *Yellow points* represent the edge pixels without coincidence. **f** Final result

Table 7.1 EMO parameters used for leukocites detection in medical images

m	n	MAXITER	δ	LISTER	I_l
50	3	5	4	4	4

Figure, the detected circle matches with only 32 pixels of the edge map. However, it is considered as the best instance due to the relationship of its internal pixels (the white pixels are much more than the black pixels). Finally, Fig. 7.5f shows final detection results over the original smear image.

Table 7.1 presents the parameters for the EMO algorithm used in this work. They have been kept for all test images after being experimentally defined.

Under such assumptions, the process to detect blood cells can thus be implemented as follows:

Step 1: Convert the RGB image to a gray scale image.
Step 2: Apply the threshold to obtain the deformed cells (first class).
Step 3: Get the edge map from the threshold image (binary image).
Step 4: Start the circle detector based in EMO over the edge map while saving best circles (Sect. 7.2.3).
Step 5: Define parameter values for each circle.

7.3   A Numerical Example of White Blood Cells Detection

In order to present the algorithm's step-by-step operation, a numerical example has been set by applying the proposed method to detect a single leukocyte lying inside of a simple image. Figure 7.6a shows the image used in the example. After applying the threshold operation, the WBC is located besides few other pixels which are merely noise (see Fig. 7.6b). Then, the edge map is subsequently computed and stored pixel by pixel inside the vector P. Figure 7.6c shows the resulting image after such procedure.

The EMO- based circle detector is executed using information of the edge map and the segmented image. Like all evolutionary approaches, EMO is a population-based optimizer that attacks the starting point problem by sampling the search space at multiple, randomly chosen, initial particles. By taking three random pixels from the vector P, three different particles are constructed. Figure 7.6d depicts the initial particle distribution. Since the particle $C_{2,0}$ holds the best fitness values $J_{New}(C_{2,0})$ (it does possess a better coincidence with the edge map and a good pixel relationship), it is considered as the best particle C^B. Then the charge of each particle is calculated according with the procedure explained in Chap. 3 and the force exerted over each particle are computed. Figure 7.6e shows the forces exerted over the $C_{3,0}$ particle. Since the $C_{3,0}$ particle is the worst in terms of the fitness value it is attracted by particles $C_{1,0}$ and $C_{2,0}$. $F_{3,1}$ and $F_{3,2}$ represent the existent attractive force of $C_{3,0}$

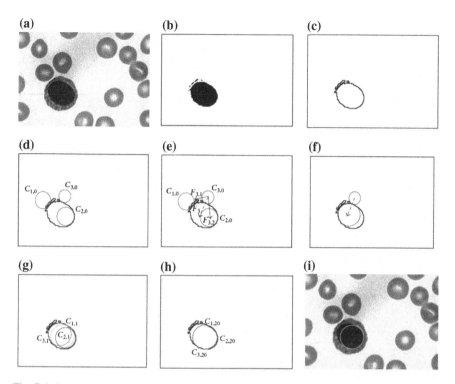

Fig. 7.6 Detection numerical example: **a** The image used as example. **b** Segmented image. **c** Edge map. **d** Initial particles. **e** Forces exerted over $C_{3,0}$. **f** New position of $C_{3,0}$. **g** Positions of all particles after the first generation. **h** Final particle configuration after 20 generations. **i** Final result overlapped the original image

with respect of $C_{1,0}$ and $C_{2,0}$ whereas F_3 corresponds to the resultant force. Considering F_3 as the final force exerted over $C_{3,0}$, the position of $C_{3,0}$ is modified according with the movement algorithm of EMO (Chap. 3). Figure 7.6f depicts the new position $C_{3,1}$ of particle $C_{3,0}$ (the second sub-index means the iteration number). If the same procedure is applied over all particles (except for $C_{2,0}$ that is the best member of the population), it yields position shown at Fig. 7.6g. Therefore after 20 iteration, all particles converge to the same position presented in Fig. 7.6h whereas the Fig. 7.6i depicts the final result.

7.4   Experimental Results

Experimental tests have been developed in order to evaluate the performance of the WBC detector. It was tested over microscope images from blood-smears holding a 600×500 pixel resolution. They correspond to supporting images on the leukemia

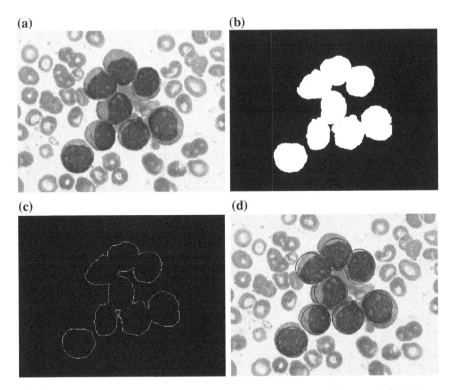

Fig. 7.7 Resulting images of the first test after applying the WBC detector: **a** Original image, **b** image segmented by the DEM algorithm, **c** edge map and **d** the white detected blood cells

diagnosis. The images show several complex conditions such as deformed cells and overlapping with partial occlusions. The robustness of the algorithm has been tested under such demanding conditions.

Figure 7.7a shows an example image employed in the test. It was used as input image for the WBC detector. Figure 7.7b presents the segmented WBC's obtained by the DEM algorithm. Figures 7.7c and d present the edge map and the white blood cells after detection, respectively. The results show that the proposed algorithm can effectively detect and mark blood cells despite cell occlusion, deformation or overlapping. Other parameters may also be calculated through the algorithm: the total area covered by white blood cells and relationships between several cell sizes.

Other example is presented in Fig. 7.8. It represents a complex example with an image showing seriously deformed cells. Despite such imperfections, the EMO algorithm can effectively detect the cells.

Figure 7.9 shows more detailed results of the detection process. From Fig. 7.9a it is possible to analyze that the blood cells are not perfect circles. Some cells are longer or have different deformations that does not fit with a perfect circle.

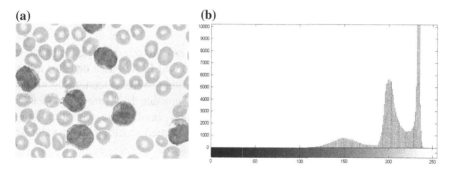

Fig. 7.8 Second test image: **a** blood cell image in gray scale, **b** its corresponding histogram

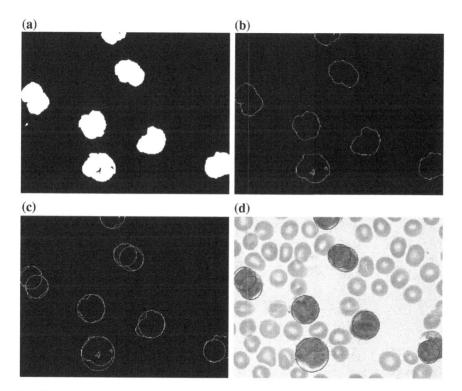

Fig. 7.9 Results after applying the circle detector based in EMO: **a** shows the binary image after the thresholding, **b** shows the edge map of the image, **c** the detected circles are drawn over the edge map image and finally in **d** the detected blood cells are enhanced

However, the results in Fig. 7.9c and d provide evidence of the detection of blood cells even whit no perfect circular shapes. The EMO based algorithm tries to find the best configuration for the entire detection of all the blood cells in the scene.

7.5   Comparisons to Other Methods

A comprehensive set of smear-blood test images is used to test the performance of the proposed approach. We have applied the proposed EMO-based detector to test images in order to compare its performance to other WBC detection algorithms such as the Boundary Support Vectors (BSV) approach [31], the iterative Otsu (IO) method [32], the Wang algorithm [33] and the Genetic algorithm-based (GAB) detector [23]. In all cases, the algorithms are tuned according to the value set which is originally proposed by their own references.

7.5.1   Detection Comparison

To evaluate the detection performance of the proposed detection method, Table 7.2 tabulates the comparative leukocyte detection performance of the BSV approach, the IO method, the Wang algorithm, the BGA detector and the proposed method, in terms of detection rates and false alarms. The experimental data set includes 30 images which are collected from the Cellavision reference library (http://www.cellavision.com). Such images contain 426 leukocytes (222 bright leukocytes and 204 dark leukocytes according to smear conditions) which have been detected and counted by a human expert. Such values act as ground truth for all the experiments. For the comparison, the detection rate (DR) is defined as the ratio between the

Table 7.2 Comparative leukocyte detection performance of the BSV approach, the IO method, the Wang algorithm, the BGA detector and the proposed EMO method over the data set which contains 30 images and 426 leukocytes

Leukocyte type	Method	Detected leukocyte	Missing	False alarms	DR (%)	FAR (%)
Bright Leukocytes (222)	BSV	104	118	67	46.85	30.18
	IO	175	47	55	78.83	24.77
	Wang	186	36	42	83.78	18.92
	BGA	177	45	22	79.73	9.91
	EMO	211	11	10	95.04	4.50
Dark Leukocytes (204)	BSV	98	106	54	48.04	26.47
	IO	166	38	49	81.37	24.02
	Wang	181	23	38	88.72	18.63
	BGA	170	134	19	83.33	9.31
	EMO	200	4	6	98.04	2.94
Overall (426)	BSV	202	224	121	47.42	28.40
	IO	341	85	104	80.05	24.41
	Wang	367	59	80	86.15	18.78
	BGA	347	79	41	81.45	9.62
	EMO	411	15	16	96.48	3.75

number of leukocytes correctly detected and the number leukocytes determined by the expert. The false alarm rate (FAR) is defined as the ratio between the number of non-leukocyte objects that have been wrongly identified as leukocytes and the number leukocytes which have been actually determined by the expert.

Experimental results show that the proposed EMO method, which achieves 96.48 % leukocyte detection accuracy with 3.75 % false alarm rate, is compared favorably against other WBC detection algorithms, such as the BSV approach, the IO method, the Wang algorithm and the BGA detector.

7.5.2 Robustness Comparison

Images of blood smear are often deteriorated by noise due to various sources of interference and other phenomena that affect the measurement processes in imaging and data acquisition systems. Therefore, the detection results depend on the algorithm's ability to cope with different kinds of noises. In order to demonstrate the robustness in the WBC detection, the proposed EMO approach is compared to the BSV approach, the IO method, the Wang algorithm and the BGA detector under noisy environments. In the test, two different experiments have been studied. The first inquest explores the performance of each algorithm when the detection task is accomplished over images corrupted by Salt and Pepper noise. The second experiment considers images polluted by Gaussian noise. Salt and Pepper and Gaussian noise are selected for the robustness analysis because they represent the most compatible noise types commonly found in images of blood smear [25, 26]. The comparison considers the complete set of 30 images presented in Sect. 7.5.1 containing 426 leukocytes which have been detected and counted by a human expert. The added noise is produced by MatLab©, considering two noise levels of 5 and 10 % for Salt and Pepper noise whereas $\sigma = 5$ and $\sigma = 10$ are used for the case of Gaussian noise. Figure 7.10 shows only two images with different noise type as

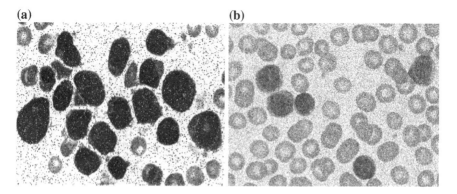

Fig. 7.10 Examples of images included in the experimental set for robustness comparison. **a** Image contaminated with 10 % of Salt and Pepper noise and **b** image polluted with $\sigma = 10$

Table 7.3 Comparative WBC detection among methods that considers the complete data set of 30 images corrupted by different levels of Salt and Pepper noise

Noise level	Method	Detected leukocyte	Missing	False alarms	DR (%)	FAR (%)
5 % Salt and Pepper noise 426 Leukocytes	BSV	148	278	114	34.74	26.76
	IO	270	156	106	63.38	24.88
	Wang	250	176	118	58.68	27.70
	BGA	306	120	103	71.83	24.18
	EMO	390	36	30	91.55	7.04
5 % Salt and Pepper noise 426 Leukocytes	BSV	101	325	120	23.71	28.17
	IO	240	186	78	56.34	18.31
	Wang	184	242	123	43.19	28.87
	BGA	294	132	83	69.01	19.48
	EMO	374	52	35	87.79	8.21

Table 7.4 Comparative WBC detection among methods that considers the complete data set of 30 images corrupted by different levels of Gaussian noise

Noise level	Method	Detected leukocyte	Missing	False alarms	DR (%)	FAR (%)
$\sigma = 5$ Gaussian noise 426 Leukocytes	BSV	172	254	77	40.37	18.07
	IO	309	117	71	72.53	16.67
	Wang	301	125	65	70.66	15.26
	BGA	345	81	61	80.98	14.32
	EMO	397	29	21	93.19	4.93
$\sigma = 10$ Gaussian noise 426 Leukocytes	BSV	143	283	106	33.57	24.88
	IO	281	145	89	65.96	20.89
	Wang	264	162	102	61.97	23.94
	BGA	308	118	85	72.30	19.95
	EMO	380	46	32	89.20	7.51

example. The outcomes in terms of the detection rate (DR) and the false alarm rate (FAR) are reported for each noise type in Tables 7.3 and 7.4. The results show that the proposed EMO algorithm presents the best detection performance, achieving in the worst case a DR of 87.79 and 89.20 %, under contaminated conditions of Salt and Pepper and Gaussian noise, respectively. On the other hand, the EMO detector possesses the least degradation performance presenting a FAR value of 8.21 and 7.51 %.

7.5.3   Stability Comparison

In order to compare the stability performance of the proposed method, its results are compared to those reported by Wang et al. in [6] which is considered as an accurate technique for the detection of WBC.

The Wang algorithm is an energy-minimizing method which is guided by internal constraint elements and influenced by external image forces, producing the segmentation of WBC's at a closed contour. As external forces, the Wang approach uses edge information which is usually represented by the gradient magnitude of the image. Therefore, the contour is attracted to pixels with large image gradients, i.e. strong edges. At each iteration, the Wang method finds a new contour configuration which minimizes the energy that corresponds to external forces and constraint elements.

In the comparison, the net structure and its operational parameters, corresponding to the Wang algorithm, follow the configuration suggested in [6] while the parameters for the EMO algorithm are taken from Table 7.1.

Figure 7.11 shows the performance of both methods considering a test image with only two white blood cells. Figure 7.11b shows that Wang's method allows

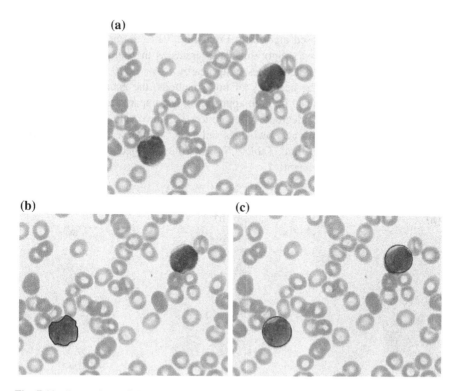

Fig. 7.11 Comparison of the EMO and the Wang's method for white blood cell detection in medical images. **a** Original image. **b** Detection using the Wang's method. **c** Detection after applying the EMO method

the detection of only one cell. Such fact can be explained by following the algorithm's evolution: it needs local and global information about the image making difficult the generalization for multiple active contours. Furthermore, Fig. 7.11c shows the result after applying the EMO method which has been proposed in this chapter.

The Wang algorithm uses the fuzzy cellular neural network (FCNN) as optimization approach. It employs gradient information and internal states in order to find a better contour configuration. In each iteration, the FCNN tries, as contour points, different new pixel positions which must be located nearby the original contour position. Such fact might cause the contour solution to remain trapped into a local minimum. In order to avoid such a problem, the Wang method applies a considerable number of iterations so that a near optimal contour configuration can be found. However, when the number of iterations increases the possibility to cover other structures increases too. Thus, if the image has a complex background (as smear images), the method gets confused so that finding the correct contour configuration from the gradient magnitude is not easy. Therefore, a drawback of Wang's method is related to its optimal iteration number (instability). Such number must be determined experimentally as it depends on the image context and its complexity. Figure 7.12a shows the result of applying 400 cycles of the Wang's algorithm while Fig. 7.12b presents the detection of the same cell shapes after 1000 iterations using the proposed algorithm. From Fig. 7.12a, it can be seen that the contour produced by Wang's algorithm degenerates as the iteration process continues, wrongly covering other shapes lying nearby.

In order to compare the accuracy of both methods, the estimated WBC area which has been approximated by both approaches, is compared to the actual WBC size considering different degrees of evolution i.e. the cycle number for each algorithm. The comparison acknowledges only one WBC for it is the only one detected by the Wang's method. Table 7.5 shows the averaged results after twenty repetitions for each experiment.

(a)   **(b)**

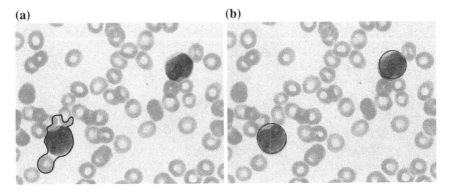

Fig. 7.12 Result comparison for the white blood cells detection showing (**a**) Wang's algorithm after 400 cycles and (**b**) EMO detector method considering 1000 cycles

Table 7.5 Error in cell's size estimation after applying the EMO algorithm and the Wang's method to detect one leukocite embedded into a blood-smear image

Algorithm	Cycles	Error (%)
Wang	60	70
	200	1
	400	121
EMO proposed	60	8.22
	200	10.1

The error is averaged after twenty experiments

7.6   Conclusions

This paper has presented an algorithm for the automatic detection of white blood cells that are embedded into complicated and cluttered smear images by considering the complete process as a circle detection problem. The approach is based on a nature-inspired technique called the Electromagnetism-Like Optimization (EMO) which is a heuristic method that follows electromagnetism principles for solving complex optimization problems. The EMO algorithm is based on electro-magnetic attraction and repulsion forces among charged particles whose charge represents the fitness solution for each particle (a given solution). The algorithm uses the encoding of three non-collinear edge points as candidate circles over an edge map. A new objective function has been derived to measure the resemblance of a candidate circle to an actual WBC based on the information from the edge map and segmentation results. Guided by the values of such objective function, the set of encoded candidate circles (charged particles) are evolved by using the EMO algorithm so that they can fit into the actual blood cells that are contained in the edge map.

The performance of the EMO-method has been compared to other existing WBC detectors (the Boundary Support Vectors (BSV) approach [31], the iterative Otsu (IO) method [32], the Wang algorithm [33] and the Genetic algorithm-based (GAB) detector [23]) considering several images which exhibit different complexity levels. Experimental results demonstrate the high performance of the proposed method in terms of detection accuracy, robustness and stability.

References

1. Liu, J., Tsui, K.: Toward nature-inspired computing. ACM Commun. **49**, 59–64 (2006)
2. Dorini, L.B., Minetto, R., Leite, N.J.: White blood cell segmentation using morphological operators and scale-space analysis. In: Proceedings of the XX Brazilian Symposium on Computer Graphics and Image Processing (2007)

3. Scotti, F.: Automatic morphological analysis for acute leukemia identification in peripheral blood microscope images. In: IEEE International Conference on Computational Intelligence for Measurement Systems and Applications, pp. 96–101 (2005)

4. Kumar, B., Joseph, D., Sreenivasc, T.: Teager energy based blood cell segmentation. In: 14th International Conference on Digital Signal Processing, pp. 619–622 (2002)

5. Cseke, I.: A fast segmentation scheme for white blood cell images. In: 11th IAPR International Conference, pp. 530–533 (2002)

6. Wang, S., Korris, F.L., Fu, D.: Applying the improved fuzzy cellular neural network IFCNN to whithe blood cell detection. Neurocomputing 70, 1348–1359 (2007)

7. Karkavitsas, G., Rangoussi, M.: Object localization in medical images using genetic algorithms. World Acad Sci Eng Technol 2, 6–9 (2005)

8. Muammar, H., Nixon, M.: Approaches to extending the Hough transform. In: Proceedings of International Conference on Acoustics, Speech and Signal Processing ICASSP-89, vol. 3, pp. 1556–1559 (1989)

9. Atherton, T.J., Kerbyson, D.J.: Using phase to represent radius in the coherent circle Hough transform. In: IEEE Colloquium on the Hough Transform, pp. 1–4 (1993)

10. Fischer, M., Bolles, R.: Random sample consensus: A paradigm to model fitting with applications to image analysis and automated cartography. CACM 24(6), 381–395 (1981)

11. Shaked, D., Yaron, O., Kiryati, N.: Deriving stopping rules for the probabilistic Hough transform by sequential analysis. Comput. Vis. Image. Underst. 63, 512–526 (1996)

12. Xu, L., Oja, E., Kultanen, P.: A new curve detection method: Randomized Hough transform (RHT). Pattern Recogn. Lett. 11(5), 331–338 (1990)

13. Han, J.H., Koczy, L.T., Poston, T.: Fuzzy Hough transform. In: Proceedings of 2nd International Conference on Fuzzy Systems, vol. 2, pp. 803–808 (1993)

14. Becker, J., Grousson, S., Coltuc, D.: From Hough transforms to integral transforms. Int. Geosci. Remote Sens. Symp. IGARSS-02. 3:1444–1446 (2002)

15. Roth, G., Levine, M.D.: Geometric primitive extraction using a genetic algorithm. IEEE Trans. Pattern Anal. 16(9), 901–905 (1994)

16. Lutton, E., Martinez, P.: A genetic algorithm for the detection 2-D geometric primitives on images. In: Proceedings of the 12th International Conference on Pattern Recognition, vol. 1, pp. 526–528 (1994)

17. Yao, J., Kharma, N., Grogono, P.: Fast robust GA-based ellipse detection. In: Proceedings of 17th International Conference on Pattern Recognition ICPR-04, vol. 2, pp. 859–862. Cambridge, UK (2004)

18. Yuen, S., Ma, C.: Genetic algorithm with competitive image labelling and least square. Pattern Recogn. 33, 1949–1966 (2000)

19. Ayala-Ramirez, V., Garcia-Capulin, C.H., Perez-Garcia, A., Sanchez-Yanez, R.E.: Circle detection on images using genetic algorithms. Pattern Recogn. Lett. 27, 652–657 (2006)

20. Cuevas, E., Oliva, D., Zaldivar, D., Pérez-Cisneros, M., Sossa, H.: Circle detection using electro-magnetism optimization. Inf. Sci. 182(1), 40–55 (2012)

21. Rosin, P.L.: Further five point fit ellipse fitting. In: Proceedings of 8th British Machine Vision Conference, pp. 290–299. Cochester, UK (1997)

22. Zhang, X., Rosin, P.L.: Superellipse fitting to partial data. Pattern Recogn. 36, 743–752 (2003)

23. Karkavitsas, G., Rangoussi, M.: Object localization in medical images using genetic algorithms. Int. J. Inf. Commun. Eng. 1(4), 204–207 (2005)

24. İlker, S., Shu-Cherng, F.: An Electromagnetism-like mechanism for global optimization. J. Global Optim. 25, 263–282 (2003)

25. Rocha, A.C., Fernandes, E.: Hybridizing the electromagnetism-like algorithm with descent search for solving engineering design problems. Int. J. Comput. Math. 86(10–11), 1932–1946 (2009)

26. Birbil, S., Fang, C., Sheu, R.L.: On the convergence of a population-based global optimization algorithm. J. Global Optim. 30(2), 301–318 (2004)

27. Bresenham, J.: A linear algorithm for incremental digital display of circular arcs. Commun. ACM **20**, 100–106 (1987)
28. Boccignone, G., Ferraro, M., Napoletano, P.: Diffused expectation maximisation for image segmentation. Electron. Lett. **40**, 1107–1108 (2004)
29. Boccignonea, G., Napoletano, P., Caggiano, V., Ferraro, M.: A multi-resolution diffused expectation–maximization algorithm for medical image segmentation. Comput. Biol. Med. **37**, 83–96 (2007)
30. Gonzalez, R.C., Woods, R.E.: Digital Image Processing. Addison Wesley, Reading (1992)
31. Wang, M., Chu, R.: A novel white blood cell detection method based on boundary support vectors. In: Proceedings of the 2009 IEEE International Conference on Systems, Man, and Cybernetics, San Antonio, TX, USA (2009)
32. Wu, J., Zeng, P., Zhou, Y., Oliver, C.: A novel color image segmentation method and its application to white blood cell image analysis. In: 8th International Conference on Signal Processing (2006)
33. Wang, S., Korris, F.L., Fu, D.: Applying the improved fuzzy cellular neural network IFCNN to white blood cell detection. Neurocomputing. **70**:1348–1359 (2007)

Chapter 8
An EMO Improvement: Opposition-Based Electromagnetism-Like for Global Optimization

8.1 Introduction

Global Optimization (GO) [1, 2] has issued applications for many areas of science [3], engineering [4], economics [5, 6] and others whose definition requires mathematical modelling [7, 8]. In general, GO aims to find the global optimum for an objective function which has been defined over a given search space. The difficulties associated with the use of mathematical methods over GO problems have contributed to the development of alternative solutions. Linear programming and dynamic programming techniques, for example, often have failed in solving (or reaching local optimum at) NP-hard problems which feature a large number of variables and non-linear objective functions. In order to overcome such problems, researchers have proposed metaheuristic-based algorithms for searching near-optimum solutions.

Metaheuristic algorithms are stochastic search methods that mimic the metaphor of biological or physical phenomena. The core of such methods lies on the analysis of collective behaviour of relatively simple agents working on decentralized systems. Such systems typically gather an agent's population that can communicate to each other while sharing a common environment. Despite a non-centralized control algorithm regulates the agent behaviour, the agent can solve complex tasks by analyzing a given global model and harvesting cooperation to other elements. Therefore, a novel global behaviour evolves from interaction among agents as it can be seen on typical examples that include ant colonies, animal herding, bird flocking, fish schooling, honey bees, bacteria, charged particles and many more. Some other metaheuristic optimization algorithms have been recently proposed to solve optimization problems, such as Genetic Algorithms (GA) [9], Particle Swarm Optimization (PSO) [10], Ant Colony Optimization (ACO) [11], Differential Evolution (DE) [12], Artificial Immune Systems (AIS) [13] and Artificial Bee Colony [14] and Gravitational Search Algorithm (GSA) [15].

© Springer International Publishing AG 2017
D. Oliva and E. Cuevas, *Advances and Applications of Optimised Algorithms in Image Processing*, Intelligent Systems Reference Library 117, DOI 10.1007/978-3-319-48550-8_8

Electromagnetism-like algorithm (EMO) is a relatively new population-based meta-heuristic algorithm which was firstly introduced by Birbil and Fang [16] to solve continuous optimization models using bounded variables. The algorithm imitates the attraction–repulsion mechanism between charged particles in an electromagnetic field. Each particle represents a solution and carries a certain amount of charge which is proportional to the solution quality (objective function). In turn, solutions are defined by position vectors which give real positions for particles within a multi-dimensional space. Moreover, objective function values of particles are calculated considering such position vectors. Each particle exerts repulsion or attraction forces over other population members; the resultant force acting over a particle is used to update its position. Clearly, the idea behind the EMO methodology is to move particles towards the optimum solution by exerting attraction or repulsion forces. Unlike other traditional meta-heuristics techniques such as GA, DE, ABC and AIS, whose population members exchange materials or information between each other, the EMO methodology assumes that each particle is influenced by all other particles in the population, mimicking other heuristics methods such as PSO and ACO. Although the EMO algorithm shares some characteristics with PSO and ACO, recent works have exhibited its better accuracy regarding optimal parameters [17–20], yet showing convergence [21]. EMO has been successfully applied to solve different sorts of engineering problems such as flow-shop scheduling [22], communications [23], vehicle routing [24], array pattern optimization in circuits [25], neural network training [26] control systems [27] and image processing [28].

EMO algorithm employs four main phases: initialization, local search, calculation and movement. The local search procedure is a stochastic search in several directions over all coordinates of each particle. EMO's main drawback is its computational complexity resulting from the large number of iterations which are commonly required during the searching process. The issue becomes worst as the dimension of the optimization problem increases. Several approaches, which simplify the local search, have been proposed in the literature to reduce EMO's computational effort. In [29] where Guan et al. proposed a discrete encoding for the particle set in order to reduce search directions at each dimension. In [30] and [31], authors include a new local search method which is based on a fixed search pattern and a shrinking strategy that aims to reduce the population size as the iterative process progresses. Additionally, in [17], a modified local search phase that employs the gradient descent method is adopted to enhance its computational complexity. Although all these approaches have improved the computational time which is required by the original EMO algorithm, recent works [27, 32] have demonstrated that reducing or simplifying EMO's local search processes also affects other important properties, such as convergence, exploration, population diversity and accuracy.

On the other hand, the opposition-based learning (OBL), that has been initially proposed in [33], is a machine intelligence strategy which considers the current estimate and its correspondent opposite value (i.e., guess and opposite guess) at the same time to achieve a fast approximation for a current candidate solution. It has been mathematically proved [34–36] that an opposite candidate solution holds a higher probability for approaching the global optimum solution than a given random candidate, yet quicker. Recently, the concept of opposition has been used to accelerate metaheuristic-based algorithms such as GA [37], DE [38], PSO [39] and GSA [40].

In this chapter, an Opposition-Based EMO called OBEMO is constructed by combining the opposition-based strategy and the standard EMO technique. The enhanced algorithm allows a significant reduction on the computational effort which required by the local search procedure yet avoiding any detriment to the good search capabilities and convergence speed of the original EMO algorithm. The presented algorithm has been experimentally tested by means of a comprehensive set of complex benchmark functions. Comparisons to the original EMO and others state-of-the-art EMO-based algorithms [7] demonstrate that the OBEMO technique is faster for all test functions, yet delivering a higher accuracy. Conclusions on the conducted experiments are supported by statistical validation that properly supports the results.

The rest of the chapter is organized as follows: Sect. 8.2 gives a simple description of OBL and Sect. 8.3 explains the implementation of the presented OBEMO algorithm. Section 8.4 presents a comparative study among OBEMO and other EMO variants over several benchmark problems. Finally, some conclusions are drawn in Sect. 8.5.

8.2 Opposition - Based Learning (OBL)

Opposition-based Learning [33] is a new concept in computational intelligence that has been employed to effectively enhance several soft computing algorithms [41, 42]. The approach simultaneously evaluates a solution x and its opposite solution $\bar{x}$ for a given problem, providing a renewed chance to find a candidate solution lying closer to the global optimum [34].

8.2.1 Opposite Number

Let $x \in [l, u]$ be a real number, where l and u are the lowest and highest bound respectively. The opposite of x is defined by:

$$\bar{x} = u + l - x \tag{8.1}$$

8.2.2 Opposite Point

Similarly, the opposite number definition is generalized to higher dimensions as follows: Let $\mathbf{x} = (x_1, x_2, \ldots, x_n)$ be a point within a n-dimensional space, where $x_1, x_2, \ldots, x_n \in R$ and $x_i \in [l_i, u_i]$, $i \in 1, 2, \ldots, n$. The opposite point $\bar{\mathbf{x}} = (\bar{x}_1, \bar{x}_2, \ldots, \bar{x}_n)$ is defined by:

$$\bar{x}_i = u_i + l_i - x_i \tag{8.2}$$

8.2.3 Opposition-Based Optimization

Metaheuristic methods start by considering some initial solutions (initial population) and trying to improve them toward some optimal solution(s). The process of searching ends when some predefined criteria are satisfied. In the absence of a priori information about the solution, random guesses are usually considered. The computation time, among others algorithm characteristics, is related to the distance of these initial guesses taken from the optimal solution. The chance of starting with a closer (fitter) solution can be enhanced by simultaneously checking the opposite solution. By doing so, the fitter one (guess or opposite guess) can be chosen as an initial solution following the fact that, according to probability theory, 50 % of the time a guess is further from the solution than its opposite guess [35]. Therefore, starting with the closer of the two guesses (as judged by their fitness values) has the potential to accelerate convergence. The same approach can be applied not only to initial solutions but also to each solution in the current population.

By applying the definition of an opposite point, the opposition-based optimization can be defined as follows: Let be $\mathbf{x}$ a point in a n-dimensional space (i.e. a candidate solution). Assume $f(\mathbf{x})$ is a fitness function which evaluates the quality of such candidate solution. According to the definition of opposite point, $\bar{\mathbf{x}}$ is the opposite of $\mathbf{x}$. If $f(\bar{\mathbf{x}})$ is better than $f(\mathbf{x})$, then $\mathbf{x}$ is updated with $\bar{\mathbf{x}}$, otherwise current point $\mathbf{x}$ is kept. Hence, the best point ($\mathbf{x}$ or $\bar{\mathbf{x}}$) is modified using known operators from the population-based algorithm.

Figure 8.1 shows the opposition-based optimization procedure. In the example, Fig. 8.1a and b represent the function to be optimized and its corresponding contour plot, respectively. By applying the OBL principles to the current population P (see Fig. 8.1b), the three particles $\mathbf{x}_1$, $\mathbf{x}_2$ and $\mathbf{x}_3$ produce a new population OP, gathering particles $\bar{\mathbf{x}}_1$, $\bar{\mathbf{x}}_2$ and $\bar{\mathbf{x}}_3$. The three fittest particles from P and OP are selected as the new population P'. It can be seen from Fig. 4b that $\bar{\mathbf{x}}_1$, $\bar{\mathbf{x}}_2$ and $\bar{\mathbf{x}}_3$ are three new members in P'. In this case, the transformation conducted on $\mathbf{x}_1$ did not provide a best chance of finding a candidate solution closer to the global optimum. Considering the OBL selection mechanism, $\bar{\mathbf{x}}_1$ is eliminated from the next generation.

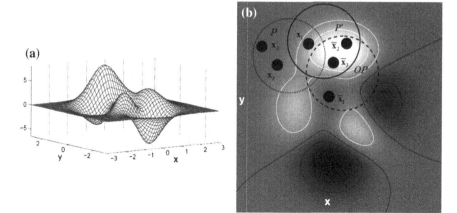

Fig. 8.1 The opposition-based optimization procedure: **a** Function to be optimized and **b** its contour plot. The current population P includes particles x_1, x_2 and x_3. The corresponding opposite population OP is represented by $\bar{x}_1$, $\bar{x}_2$ and $\bar{x}_3$. The final population P' is obtained by the OBL selection mechanism yielding particles $\bar{x}_1$, $\bar{x}_2$ and $\bar{x}_3$

8.3 Opposition-Based Electromagnetism-Like Optimization (OBEMO)

Similarly to all metaheuristic-based optimization algorithms, two steps are fundamental for the EMO algorithm: the population initialization and the production of new generations by evolutionary operators. In the approach, the OBL scheme is incorporated to enhance both steps. However, the original EMO is considered as the main algorithm while the opposition procedures are embedded into EMO aiming to accelerate its convergence speed. Figure 8.2 shows a data flow comparison between the EMO and the OBEMO algorithm. The novel extended opposition procedures are explained in the following subsections.

8.3.1 Opposition-Based Population Initialization

In population-based meta-heuristic techniques, the random number generation is the common choice to create an initial population in absence of a priori knowledge. Therefore, as mentioned in Sect. 8.2, it is possible to obtain fitter starting candidate solutions by utilizing OBL despite no a priori knowledge about the solution(s) is available. The following steps explain the overall procedure.

(1) Initialize the population $\mathbf{X}$ with N_P representing the number of particles.
(2) Calculate the opposite population by

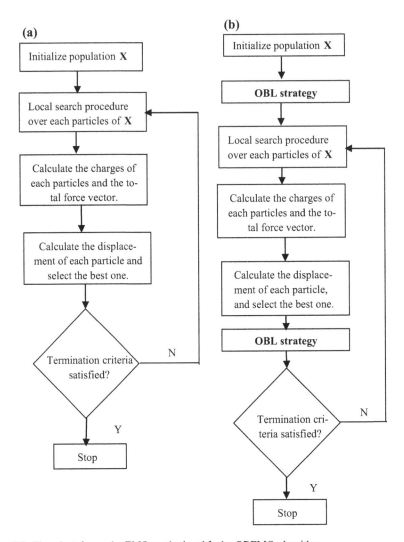

Fig. 8.2 Flowchart for: **a** the EMO method and **b** the OBEMO algorithm

$$\bar{x}_i^j = u_i + l_i - x_i^j$$
$$i = 1, 2, \ldots, n; j = 1, 2, \ldots, N_P$$ (8.3)

where x_i^j and $\bar{x}_i^j$ denote the ith parameter of the jth particle of the population and its corresponding opposite particle.

(3) Select the N_P fittest elements from $\{\mathbf{X} \cup \bar{\mathbf{X}}\}$ as initial population.

8.3.2 Opposition-Based Production for New Generation

Starting from the current population, the OBL strategy can be used again to produce new populations. In this procedure, the opposite population is calculated and the fittest individuals are selected from the union of the current population and the opposite population. The following steps summarize the OBEMO implementation as follows:

Step 1: Generate N_P initial random particles $\mathbf{x}^h$ to create the particle vector $\mathbf{X}$, with $h \in 1, 2, \ldots N_P$.

Step 2: Apply the OBL strategy by considering N_P particles from vector $\mathbf{X}$ and generating the opposite vector $\overline{\mathbf{X}}$ through Eq. 8.2.

Step 3: Select the N_P fittest particles from $\mathbf{X} \cup \overline{\mathbf{X}}$ according to $f(\cdot)$. These particles build the initial population $\mathbf{X}_0$.

Step 4: Calculate the local search procedure for each particle of $\mathbf{X}_0$ as follows: For a given dimension d, the particle $\mathbf{x}^h$ is assigned to a temporary point y to store the initial information. Next, a random number is selected and combined with δ to yield the step length. Therefore, the point y is moved along that direction. The sign is determined randomly. If $f(\mathbf{x}^h)$ is minimized, the particle $\mathbf{x}^h$ is replaced by y, ending the neighborhood-wide search for a particle h. The result is stored into the population vector $\mathbf{X}_{Local}$.

Step 5: Determine the best particle $\mathbf{x}^{best}$ of the population vector $\mathbf{X}_{Local}$ (with $\mathbf{x}^{best} \leftarrow \arg\min\{f(\mathbf{x}^h), \forall h\}$).

Step 6: Calculate the charge among particles and the vector force through according the EMO procedures presented in Chap. 3. The particle showing the better objective function value holds a bigger charge and therefore a bigger attraction force.

Step 7: Change particle positions according to their force magnitude. The new particle's position is calculated (see Chap. 3). $\mathbf{x}^{best}$ is not moved because it has the biggest force and attracts others particles to itself. The result is stored into the population vector $\mathbf{X}_{Mov}$.

Step 8: Apply the OBL strategy over the m particles of the population vector $\mathbf{X}_{Mov}$, the opposite vector $\overline{\mathbf{X}}_{Mov}$ can be calculated through Eq. 8.2.

Step 9: Select the m fittest particles from $\mathbf{X}_{Mov} \cup \overline{\mathbf{X}}_{Mov}$ according to $f(\cdot)$. Such particles represent the population $\mathbf{X}_0$.

Step 10: Increase the *Iteration* index. If *iteration* = *MAXITER or the* value of $f(X)$ is smaller than the pre-defined threshold value, then the algorithm is stopped and the flow jumps to step 11. Otherwise, it jumps to step 4.

Step 11: The best particle $\mathbf{x}^{best}$ is selected from the last iteration as it is considered as the solution.

8.4   Experimental Results

In order to test the algorithm's performance, the presented OBEMO is compared to the standard EMO and others state-of-the-art EMO-based algorithms. In this section, the experimental results are discussed in the following subsections:

- Test problems
- Parameter settings for the involved EMO algorithms
- Results and discussions

8.4.1   Test Problems

A comprehensive set of benchmark problems, that includes 14 different global optimization tests, has been chosen for the experimental study. According to their use in the performance analysis, the functions are divided in two different sets: original test functions $(f_1 - f_9)$ and multidimensional functions $(f_{10} - f_{14})$. Every function at this paper is considered as a minimization problem itself.

The original test functions, which are shown in Table 8.1, agree to the set of numerical benchmark functions presented by the original EMO paper at [16]. Considering that such function set is also employed by a vast majority of EMO-based new approaches, its use in our experimental study facilitates its comparison to similar works. More details can be found in [43].

The major challenge of an EMO-based approach is to avoid the computational complexity that arises from the large number of iterations which are required during the local search process. Since the computational complexity depends on the dimension of the optimization problem, one set of multidimensional functions (see Table 8.2) is used in order to assess the convergence and accuracy for each algorithm. Multidimensional functions include a set of five different functions whose dimension has been fixed to 30.

8.4.2   Parameter Settings for the Involved EMO Algorithms

The experimental set aims to compare four EMO-based algorithms including the presented OBEMO. All algorithms face 14 benchmark problems. The algorithms are listed below:

- Standard EMO algorithm [16];
- Hybridizing EMO with descent search (HEMO) [17];
- EMO with fixed search pattern (FEMO) [30];
- The presented approach OBEMO.

Table 8.1 Optimization test functions corresponding to the original test set

Function	Search domain	Global minima
Branin $f_1(x_1,x_2) = \left(x_2 - \frac{5}{4\pi^2}x_1^2 + \frac{5}{\pi}x_1 - 6\right)^2 + 10\left(1 - \frac{1}{8\pi}\right)\cos x_1 + 10$	$-5 \leq x_1 \leq 10$ $0 \leq x_2 \leq 15$	0.397887
Camel $f_2(x_1,x_2) = -\dfrac{-x_1^2 + 4.5x_1^2 + 2}{e^{2x_2^2}}$	$-2 \leq x_1, x_2 \leq 2$	-1.031
Goldenstain-Price $f_3(x_1,x_2) = 1 + (x_1 + x_2 + 1)^2 \times (19 - 14x_1 + 13x_1^2 - 14x_2 + 6x_1x_2 + 3x_2^2)$ $\times (30 + 2x_1 - 3x_2)^2 \times (18 - 32x_1 + 12x_1^2 - 48x_2 - 36x_1x_2 + 27x_2^2)$	$-2 \leq x_1, x_2 \leq 2$	3.0
Hartmann (3-dimensional) $f_4(\mathbf{x}) = -\sum\limits_{i=1}^{4} \alpha_i \exp\left[-\sum\limits_{j=1}^{3} A_{ij}(x_j - P_{ij})^2\right]$ $\alpha = [1, 1.2, 3, 3.2],\ \mathbf{A} = \begin{bmatrix} 3.0 & 10 & 30 \\ 0.1 & 10 & 35 \\ 3.0 & 10 & 30 \\ 0.1 & 10 & 35 \end{bmatrix},\ \mathbf{P} = 10^{-4}\begin{bmatrix} 6890 & 1170 & 2673 \\ 4699 & 4387 & 7470 \\ 1091 & 8732 & 5547 \\ 381 & 5743 & 8828 \end{bmatrix}$	$0 \leq x_i \leq 1$ $i = 1, 2, 3$	-3.8627
Hartmann (6-dimensional) $f_5(\mathbf{x}) = -\sum\limits_{i=1}^{4} \alpha_i \exp\left[-\sum\limits_{j=1}^{6} B_{ij}(x_j - Q_{ij})^2\right]$ $\alpha = [1, 1.2, 3, 3.2],\ \mathbf{B} = \begin{bmatrix} 10 & 3 & 17 & 3.05 & 1.7 & 8 \\ 0.05 & 10 & 17 & 0.1 & 8 & 14 \\ 3 & 3.5 & 1.7 & 10 & 17 & 8 \\ 17 & 8 & 0.05 & 10 & 0.1 & 14 \end{bmatrix},$ $\mathbf{Q} = 10^{-4}\begin{bmatrix} 1312 & 1696 & 5569 & 124 & 8283 & 5886 \\ 2329 & 4135 & 8307 & 3736 & 1004 & 9991 \\ 2348 & 1451 & 3522 & 2883 & 3047 & 6650 \\ 4047 & 8828 & 8732 & 5743 & 1091 & 381 \end{bmatrix}$	$0 \leq x_i \leq 1$ $i = 1, 2, 3, \ldots, 6$	-3.8623

(continued)

Table 8.1 (continued)

Function	Search domain	Global minima
Shekel S_m (4-dimensional) $S_m(\mathbf{x}) = -\sum_{j=1}^{m}\left[\sum_{i=1}^{4}(x_i - C_{ij})^2 + \beta_j\right]^{-1}$ $\boldsymbol{\beta} = [1, 2, 2, 4, 4, 6, 3, 7, 5, 5]^T,$ $\mathbf{C} = \begin{bmatrix} 4.0 & 1.0 & 8.0 & 6.0 & 3.0 & 2.0 & 5.0 & 8.0 & 6.0 & 7.0 \\ 4.0 & 1.0 & 8.0 & 6.0 & 7.0 & 9.0 & 5.0 & 1.0 & 2.0 & 3.6 \\ 4.0 & 1.0 & 8.0 & 6.0 & 3.0 & 2.0 & 3.0 & 8.0 & 6.0 & 7.0 \\ 4.0 & 1.0 & 8.0 & 6.0 & 7.0 & 9.0 & 3.0 & 1.0 & 2.0 & 3.6 \end{bmatrix}$	$0 \leq x_i \leq 1$ $i = 1, 2, 3, 4$	
$f_6(\mathbf{x}) = S_5(\mathbf{x})$		-10.1532
$f_7(\mathbf{x}) = S_7(\mathbf{x})$		-10.4029
$f_8(\mathbf{x}) = S_{10}(\mathbf{x})$		-10.5364
Shubert $f_9(x_1, x_2) = \left(\sum_{i=1}^{5} i\cos((i+1)x_1 + i)\right)\left(\sum_{i=1}^{5} i\cos((i+1)x_2 + i)\right)$	$-10 \leq x_1, x_2 \leq 10$	-186.73

Table 8.2 Multidimensional test function set

Function	Search domain	Global minima
$f_{10}(\mathbf{x}) = \sum_{i=1}^{n} \left[x_i^2 - 10\cos(2\pi x_i) + 10 \right]$	$[-5.12, 5.12]^{30}$	0
$f_{11}(\mathbf{x}) = -20\exp\left(-0.2\sqrt{\frac{1}{n}\sum_{i=1}^{n} x_i^2}\right) - \exp\left(\frac{1}{n}\sum_{i=1}^{n}\cos(2\pi x_i)\right) + 20$	$[-32, 32]^{30}$	0
$f_{12}(\mathbf{x}) = \frac{1}{4000}\sum_{i=1}^{n} x_i^2 - \prod_{i=1}^{n}\cos\left(\frac{x_i}{\sqrt{i}}\right) + 1$	$[-600, 600]^{30}$	0
$f_{13}(\mathbf{x}) = \frac{\pi}{n}\left\{ 10\sin(\pi y_1) + \sum_{i=1}^{n-1}(y_i - 1)^2\left[1 + 10\sin^2(\pi y_{i+1})\right] + (y_n - 1)^2 \right\}$ $+ \sum_{i=1}^{n} u(x_i, 10, 100, 4)$ $u(x_i, a, k, m) = \begin{cases} k(x_i - a)^m & x_i > a \\ 0 & -a < x_i < a \\ k(-x_i - a)^m & x_i < -a \end{cases}$	$[-50, 50]^{30}$	0
$f_{14}(\mathbf{x}) = \sin^2(3\pi x_1) + \sum_{i=1}^{n}(x_i - 1)^2\left[1 + \sin^2(3\pi x_i + 1)\right]$ $+ (x_n - 1)^2\left[1 + \sin^2(2\pi x_n)\right] + \sum_{i=1}^{n} u(x_i, 5, 100, 4)$	$[-50, 50]^{30}$	0

For the original EMO algorithm described in [16] and the proposed OBEMO, the parameter set is configured considering: $\delta = 0.001$ and *LISTER* = 4. For the HEMO, the following experimental parameters are considered: $LsIt_{\max} = 10$, $\varepsilon_r = 0.001$ and $\gamma = 0.00001$. Such values can be assumed as the best configuration set according to [17]. Diverging from the standard EMO and the OBEMO algorithm, the HEMO method reduces the local search phase by only processing the best found particle $\mathbf{x}^{best}$. The parameter set for the FEMO approach is defined by considering the following values: $N_{fe}^{\max} = 100$, $N_{ls}^{\max} = 10$, $\delta = 0.001$, $\delta^{\min} = 1 \times 10^{-8}$ and $\varepsilon_\delta = 0.1$. All aforementioned EMO-based algorithms use the same population size of $m = 50$.

8.4.3   Results and Discussions

Original test functions set

On this test set, the performance of the OBEMO algorithm is compared to standard EMO, HEMO and FEMO, considering the original test functions set. Such functions, presented in Table 8.1, hold different dimensions and one known global minimum. The performance is analyzed by considering 35 different executions for each algorithm. The case of no significant changes in the solution being registered (i.e. smaller than 10^{-4}) is considered as stopping criterion.

The results, shown by Table 8.3, are evaluated assuming the averaged best value $f(x)$ and the averaged number of executed iterations (*MAXITER*). Figure 8.3 shows the optimization process for the function f_3 and f_6. Such function values correspond to the best case for each approach that is obtained after 35 executions.

In order to statistically analyse the results in Table 8.3, a non-parametric significance proof known as the Wilcoxon's rank test [44–46] has been conducted. Such proof allows assessing result differences among two related methods. The analysis is performed considering a 5 % significance level over the "averaged best value of $f(x)$" and the "averaged number of executed iterations of *MAXITER*" data. Tables 8.4 and 8.5 reports the p-values produced by Wilcoxon's test for the pair-wise comparison of the "averaged best value" and the "averaged number of executed iterations" respectively, considering three groups. Such groups are formed by OBEMO versus EMO, OBEMO versus HEMO and OBEMO versus FEMO. As a null hypothesis, it is assumed that there is no difference between the values of the two algorithms. The alternative hypothesis considers an actual difference between values from both approaches. The results obtained by the Wilcoxon test indicate that data cannot be assumed as occurring by coincidence (i.e. due to the normal noise contained in the process).

Table 8.4 considers the Wilcoxon analysis with respect to the "averaged best value" of $f(x)$. The p-values for the case of OBEMO versus EMO are larger than 0.05 (5 % significance level) which is a strong evidence supporting the null hypothesis which indicates that there is no significant difference between both

Table 8.3 Comparative results for the EMO, the OBEMO, the HEMO and the FEMO algorithms considering the original test functions set (Table 1)

Function		f_1	f_2	f_3	f_4	f_5	f_6	f_7	f_8	f_9
Dimension		2	2	2	3	6	4	4	4	2
EMO	Averaged best values $f(x)$	0.3980	−1.015	3.0123	−3.7156	−3.6322	−10.07	−10.23	−10.47	−186.71
	Averaged *MAXITER*	103	128	197	1.59E + 03	1.08E + 03	30	31	29	44
OBEM	Averaged best values $f(x)$	0.3980	−1.027	3.0130	−3.7821	−3.8121	−10.11	−10.22	−10.50	−186.65
	Averaged *MAXITER*	61	83	101	1.12E + 03	826	18	19	17	21
HEMO	Averaged best values $f(x)$	0.5151	−0.872	3.413	−3.1187	−3.0632	−9.041	−9.22	−9.1068	−184.31
	Averaged *MAXITER*	58	79	105	1.10E + 03	805	17	18	15	22
FEMO	Averaged best values $f(x)$	0.4189	−0.913	3.337	−3.3995	−3.2276	−9.229	−9.88	−10.18	−183.88
	Averaged *MAXITER*	63	88	98	1.11E + 03	841	21	22	19	25

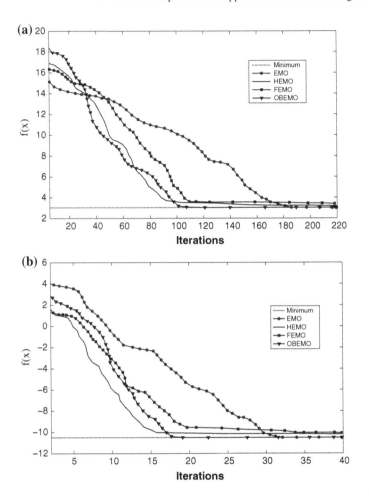

Fig. 8.3 Comparison of the optimization process for two original test functions: **a** f_3 and **b** f_6

Table 8.4 Results from Wilcoxon's ranking test considering the "averaged best value of $f(x)$"

Function	p-Values		
	OBEMO versus EMO	OBEMO versus HEMO	OBEMO versus FEMO
f_1	0.3521	1.21E-04	1.02E-04
f_2	0.4237	1.05E-04	0.88E-04
f_3	0.2189	4.84E-05	3.12E-05
f_4	0.4321	1.35E-05	1.09E-05
f_5	0.5281	2.73E-04	2.21E-04
f_6	0.4219	1.07E-04	0.77E-04
f_7	0.3281	3.12E-05	2.45E-05
f_8	0.4209	4.01E-05	3.62E-05
f_9	0.2135	1.86E-05	1.29E-05

Table 8.5 Results from Wilcoxon's ranking test considering the "averaged number of executed iterations"

Function	p-Values		
	OBEMO versus EMO	OBEMO versus HEMO	OBEMO versus FEMO
f_1	2.97E-04	0.2122	0.2877
f_2	3.39E-04	0.1802	0.2298
f_3	8.64E-09	0.1222	0.1567
f_4	7.54E-05	0.2183	0.1988
f_5	1.70E-04	0.3712	0.3319
f_6	5.40E-13	0.4129	0.3831
f_7	7.56E-04	0.3211	0.3565
f_8	1.97E-04	0.2997	0.2586
c	1.34E-05	0.3521	0.4011

methods. On the other hand, in cases for the p-values corresponding to the OBEMO versus HEMO and OBEMO versus FEMO, they are less than 0.05 (5 % significance level), which accounts for a significant difference between the "averaged best value" data among methods. Table 8.5 considers the Wilcoxon analysis with respect to the "averaged number of executed iterations" values. Applying the same criteria, it is evident that there is a significant difference between the OBEMO versus EMO case, despite the OBEMO versus HEMO and OBEMO versus FEMO cases offering similar results.

Multidimensional functions

In contrast to the original functions, Multidimensional functions exhibit many local minima/maxima which are, in general, more difficult to optimize. In this section the performance of the OBEMO algorithm is compared to the EMO, the HEMO and the FEMO algorithms, considering functions in Table 8.2. This comparison reflects

Table 8.6 Comparative results for the EMO, OBEMO, HEMO and the FEMO algorithms being applied to the multidimensional test functions (Table 8.2)

Function		f_{10}	f_{11}	f_{12}	f_{13}	f_{14}
Dimension		30	30	30	30	30
EMO	Averaged best values $f(x)$	2.12E-05	1.21E-06	1.87E-05	1.97E-05	2.11E-06
	Averaged *MAXITER*	622	789	754	802	833
OBEM	Averaged best values $f(x)$	3.76E-05	5.88E-06	3.31E-05	4.63E-05	3.331E-06
	Averaged *MAXITER*	222	321	279	321	342
HEMO	Averaged best values $f(x)$	2.47E-02	1.05E-02	2.77E-02	3.08E-02	1.88E-2
	Averaged *MAXITER*	210	309	263	307	328
FEMO	Averaged best values $f(x)$	1.36E-02	2.62E-02	1.93E-02	2.75E-02	2.33E-02
	Averaged *MAXITER*	241	361	294	318	353

the algorithm's ability to escape from poor local optima and to locate a near-global optimum, consuming the least number of iterations. The dimension of such functions is set to 30. The results (Table 8.6) are averaged over 35 runs reporting the "averaged best value" and the "averaged number of executed iterations" as performance indexes.

Table 8.7 Results from Wilcoxon's ranking test considering the "best averaged values"

Function	p-Values		
	OBEMO versus EMO	OBEMO versus HEMO	OBEMO versus FEMO
f_{10}	0.2132	3.21E-05	3.14E-05
f_{11}	0.3161	2.39E-05	2.77E-05
f_{12}	0.4192	5.11E-05	1.23E-05
f_{13}	0.3328	3.33E-05	3.21E-05
f_{14}	0.4210	4.61E-05	1.88E-05

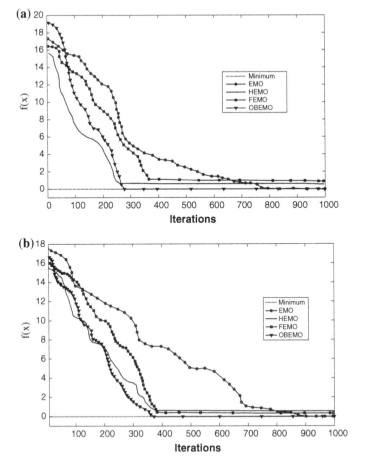

Fig. 8.4 Optimization process comparison for two multidimensional test functions: **a** f_{12} and **b** f_{14}

Table 8.8 Results from Wilcoxon's ranking test considering the "averaged number of executed iterations"

Function	p-Values		
	OBEMO versus EMO	OBEMO versus HEMO	OBEMO versus FEMO
f_{10}	3.78E-05	0.1322	0.2356
f_{11}	2.55E-05	0.2461	0.1492
f_{12}	6.72E-05	0.3351	0.3147
f_{13}	4.27E-05	0.2792	0.2735
f_{14}	3.45E-05	0.3248	0.3811

The Wilcoxon rank test results, presented in Table 8.7, shows that the p-values (regarding to the "averaged best value" values of Table 8.6) for the case of OBEMO versus EMO, indicating that there is no significant difference between both methods. p-values corresponding to the OBEMO versuss HEMO and OBEMO versus FEMO show that there is a significant difference between the "averaged best" values among the methods. Figure 8.4 shows the optimization process for the function. Such function values correspond to the best case, for each approach, obtained after 35 executions.

Table 8.8 considers the Wilcoxon analysis with respect to the "averaged number of executed iterations" values of Table 8.6. As it is observed, the outcome is similar to the results from last test on the original functions.

8.5 Conclusions

In this chapter, an Opposition-Based EMO, named as OBEMO, has been presented by combining the opposition-based learning (OBL) strategy and the standard EMO technique. The OBL is a machine intelligence strategy which considers, at the same time, a current estimate and its opposite value to achieve a fast approximation for a given candidate solution. The standard EMO is enhanced by using two OBL steps: the population initialization and the production of new generations. The enhanced algorithm significantly reduces the required computational effort yet avoiding any detriment to the good search capabilities of the original EMO algorithm.

A set of 14 benchmark test functions has been employed for experimental study. Results are supported by a statistically significant framework (Wilcoxon test [44–46]) to demonstrate that the OBEMO is as accurate as the standard EMO yet requiring a shorter number of iterations. Likewise, it is as fast as others state-of-the-art EMO-based algorithms such as HEMO [7] and FEMO [30], still keeping the original accuracy.

Although the results offer evidence to demonstrate that the Opposition-Based EMO method can yield good results on complicated optimization problems, the chapter's aim is not to devise an optimization algorithm that could beat all others currently available, but to show that the Opposition-based Electromagnetism-like method can effectively be considered as an attractive alternative for solving global optimization problems.

References

1. Tan, S., Cheng, X., Hongbo, X.: An efficient global optimization approach for rough set based dimensionality reduction. Int. J. Innovative Comput. Inf. Control **3**(3), 725–736 (2007)
2. Borji, A., Hamidi, M.: A new approach to global optimization motivated by parliamentary political competitions. Int. J. Innovative Comput. Inf. Control **5**(6), 1643–1653 (2009)
3. Yang, C.N., Huang, K.S., Yang, C.B., Hsu, C.Y.: Error-tolerant minimum finding with DNA computing. Int. J. Innovative Comput. Inf. Control. **5**(10(A)), 3045–3057 (2009)
4. Gao, W., Ren, H.: An optimization model based decision support system for distributed energy systems planning. Int. J. Innovative Comput. Inf. Control. **7**(5(B)), pp. 2651–2668 (2011)
5. Chunhui, X., Wang, J., Shiba, N.: Multistage portfolio optimization with var as risk measure. Int. J. Innovative Comput. Inf. Control **3**(3), 709–724 (2007)
6. Chang, J.F.:. A performance comparison between genetic algorithms and particle swarm optimization applied in constructing equity portfolios. Int. J. Innovative Comput. Inf. Control. **5**(12(B)), pp. 5069–5079 (2009)
7. Takeuchi, Y.: Optimization of linear observations for the stationary kalman filter based on a generalized water filling theorem. Int. J. Innovative Comput. Inf. Control **4**(1), 211–230 (2008)
8. Borzabadi, A.H., Sadjadi, M.E., Moshiri, B.: A numerical scheme for approximate optimal control of nonlinear hybrid systems. Int. J. Innovative Comput. Inf. Control **6**(6), 2715–2724 (2010)
9. Holland, J.H.: Adaptation in Natural and Artificial Systems. University of Michigan Press, Ann Arbor (1975)
10. Kennedy, J., Eberhart, R.: Particle swarm optimization. In: IEEE International Conference on Neural Networks (Piscataway, NJ), pp. 1942–1948 (1995)
11. Dorigo, M., Maniezzo, V., Colorni, A.: Positive feedback as a search strategy, Technical Report 91-016. Politecnico di Milano, Italy (1991)
12. Price, K., Storn, R., Lampinen, A.: Differential Evolution a Practical Approach to Global Optimization. Springer Natural Computing Series, Berlin (2005)
13. Fyfe, C., Jain, L.: Teams of intelligent agents which learn using artificial immune systems. J Network Comput. Appl. **29**(2–3), 147–159 (2005)
14. Karaboga, D.: An idea based on honey bee swarm for numerical optimization, technical report-TR06,Erciyes University, Engineering Faculty, Computer Engineering Department (2005)
15. Rashedia, E., Nezamabadi-pour, H., Saryazdi, S.: Filter modeling using Gravitational Search Algorithm. Eng. Appl. Artif. Intell. **24**(1), 117–122 (2011)
16. İlker, S.: Birbil and Shu-Cherng Fang. An electromagnetism-like mechanism for global optimization. J. Global Optim. **25**, 263–282 (2003)
17. Rocha, A., Fernandes, E.: Hybridizing the electromagnetism-like algorithm with descent search for solving engineering design problems. Int. J. Comput. Math. **86**, 1932–1946 (2009)
18. Rocha, A., Fernandes, E.: Modified movement force vector in an electromagnetism-like mechanism for global optimization. Optim. Methods Softw. **24**, 253–270 (2009)
19. Tsou, C.S., Kao, C.H.: Multi-objective inventory control using electromagnetism-like metaheuristic. Int. J. Prod. Res. **46**, 3859–3874 (2008)

20. Wu, P., Wen-Hung, Y., Nai-Chieh, W.: An electromagnetism algorithm of neural network analysis an application to textile retail operation. J. Chin. Inst. Ind. Eng. **21**, 59–67 (2004)
21. Birbil, S.I., Fang, S.C., Sheu, R.L.: On the convergence of a population-based global optimization algorithm. J. Global Optim. **30**(2), 301–318 (2004)
22. Naderi, B., Tavakkoli-Moghaddam, R., Khalili, M.: Electromagnetism-like mechanism and simulated annealing algorithms for flowshop scheduling problems minimizing the total weighted tardiness and makespan. Knowl. Based Syst. **23**, 77–85 (2010)
23. Hung H.L., Huang, Y.F.: Peak to average power ratio reduction of multicarrier transmission systems using electromagnetism-like method. Int. J. Innovative Comput. Inf. Control. **7**(5(A)), 2037–2050 (2011)
24. Yurtkuran, A., Emel, E.: A new hybrid electromagnetism-like algorithm for capacitated vehicle routing problems. Expert Syst. Appl. **37**, 3427–3433 (2010)
25. Jhen-Yan, J., Kun-Chou, L.: Array pattern optimization using electromagnetism-like algorithm. AEU Int. J. Electron. Commun. **63**, 491–496 (2009)
26. Wu, P., Wen-Hung, Y., Nai-Chieh, W.: An electromagnetism algorithm of neural network analysis an application to textile retail operation. J. Chin. Inst. Ind. Eng. **21**, 59–67 (2004)
27. Lee, C.H., Chang, F.K.: Fractional-order PID controller optimization via improved electromagnetism-like algorithm. Expert Syst. Appl. **37**, 8871–8878 (2010)
28. Cuevas, E., Oliva, D., Zaldivar, D., Pérez-Cisneros, M., Sossa, H.: Circle detection using electro-magnetism optimization. Inf. Sci. **182**(1), 40–55 (2012)
29. Guan, X., Dai, X., Li, J.: Revised electromagnetism-like mechanism for flow path design of unidirectional AGV systems. Int. J. Prod. Res. **49**(2), 401–429 (2011)
30. Rocha, A.M.A.C., Fernandes, E.M.G.P.: Numerical experiments with a population shrinking strategy within a electromagnetism-like algorithm. J. Math. Comput Simul. **1**(3), 238–243 (2007)
31. Rocha, A.M.A.C., Fernandes, E.M.G.P.: Numerical study of augmented Lagrangian algorithms for constrained global optimization. Optimization. **60**(10–11), 1359–1378 (2011)
32. Lee, C.H., Chang, F.K., Kuo, C.T., Chang, H.H.: A hybrid of electromagnetism-like mechanism and back-propagation algorithms for recurrent neural fuzzy systems design. Int. J. Syst. Sci. **43**(2), 231–247 (2012)
33. Tizhoosh, H.R.: Opposition-based learning: a new scheme for machine intelligence. In: Proceedings of International Conference on Computational Intelligence for Modeling Control and Automation, pp. 695–701 (2005)
34. Rahnamayn, S., Tizhoosh, H.R., Salama, M.: A novel population initialization method for accelerating evolutionary algorithms. Comput. Math Appl. **53**(10), 1605–1614 (2007)
35. Rahnamayan, S., Tizhoosh, H.R., Salama, M.M.A.: Opposition versus randomness in soft computing techniques. Elsevier J. Appl. Soft Comput. **8**, 906–918 (2008)
36. Wang, H., Zhijian, W., Rahnamayan, S.: Enhanced opposition-based differential evolution for solving high-dimensional continuous optimization problems. Soft. Comput. (2010). doi:10. 1007/s00500-010-0642-7
37. Iqbal, M.A., Khan, N.K., Multaba, H., Rauf Baig, A.: A novel function optimization approach using opposition based genetic algorithm with gene excitation. Int. J. Innovative Comput. Inf. Control. **7**(7(B)), 4263–4276 (2011)
38. Rahnamayan, S., Tizhoosh, H.R., Salama, M.M.A.: Opposition-based differential evolution. IEEE Trans. Evol. Comput. **12**(1), 64–79 (2008)
39. Wanga, H., Wua, Z., Rahnamayan, S., Liu, Y., Ventresca, M.: Enhancing particle swarm optimization using generalized opposition-based learning. Inf. Sci. **181**, 4699–4714 (2011)
40. Shaw, B., Mukherjee, V., Ghoshal, S.P.: A novel opposition-based gravitational search algorithm for combined economic and emission dispatch problems of power systems. Electr. Power Energy Syst. **35**, 21–33 (2012)
41. Tizhoosh, H.R.: Opposition-based reinforcement learning. J. Adv. Comput. Intell. Intell. Inform. **10**(3), 578–585 (2006)
42. Shokri, M., Tizhoosh, H.R., Kamel, M.: Opposition-based Q(k) algorithm. In: Proceedings of IEEE World Congress Computational Intelligence, pp. 646–53 (2006)

43. Dixon, L.C.W., Szegö, G.P.: The global optimization problem: An introduction. Towards Global Optimization 2, North-Holland, Amsterdam, pp. 1–15 (1978)
44. Wilcoxon, F.: Individual comparisons by ranking methods. Biometrics **1**, 80–83 (1945)
45. Garcia, S., Molina, D., Lozano, M., Herrera, F.: A study on the use of non-parametric tests for analyzing the evolutionary algorithms' behaviour: A case study on the CEC'2005 Special session on real parameter optimization. J. Heurist (2008). doi:10.1007/s10732-008-9080-4
46. Santamaría, J., Cordón, O., Damas, S., García-Torres, J.M., Quirin, A.: Performance evaluation of memetic approaches in 3D reconstruction of forensic objects. Soft Comput. doi:10.1007/s00500-008-0351-7, in press (2008)

Printed in the United States
By Bookmasters